a spacefaring people

NASA SP-4405

a spacefaring people:

PERSPECTIVES ON EARLY SPACEFLIGHT

Alex Roland

The NASA History Series

NASA Scientific and Technical Information Branch   1985
National Aeronautics and Space Administration
Washington, DC

NASA maintains an internal history program for two principal reasons: (1) Sponsorship of research in NASA-related history is one way in which NASA responds to the provision of the National Aeronautics and Space Act of 1958 that requires NASA to "provide for the widest practicable and appropriate dissemination of information concerning its activities and the results thereof." (2) Thoughtful study of NASA history can help agency managers accomplish the missions assigned to the agency. Understanding NASA's past aids in understanding its present situation and illuminates possible future directions. The opinions and conclusions set forth in this book are those of the authors; no official of the agency necessarily endorses those opinions or conclusions.

Library of Congress Cataloging in Publication Data
Main entry under title:
A spacefaring people.
   (NASA SP; 4405)
   "Essays . . . delivered at a conference on the history of space activity held at Yale University on February 6 and 7, 1981"—Introd.
   1. Astronautics—History—Congresses. I. Roland, Alex, 1944-
TL788.5.S64   1985   629.4'09   84-979

For sale by the Superintendent of Documents, U.S. Government Printing Office
Washington, D.C. 20402

CONTENTS

CONTENTS

INTRODUCTION

The essays on the early years of spaceflight that follow were originally presented at a conference on the history of space activity, held at Yale University on February 6 and 7, 1981. The conference grew out of a course I offered at Yale University in the fall of 1980 entitled "NASA and the Post-Sputnik Era." Jointly sponsored by Calhoun and Jonathan Edwards colleges in response to student interest, the course was quickly oversubscribed. Therefore, the first purpose of this conference was to provide a larger forum in which Yale students could observe—and participate in—informed discussions about United States space activity to date.

The conference attracted a far wider and more diverse audience than expected. People from all over the country came to New Haven in a month when that city is not at its best, and participated actively in the conference and all activities related to it.

These proceedings would no doubt be richer and more representative of the conference had it been possible to capture and transcribe much of the discussions about the formal papers. Papers included here are basically in the same form as originally presented, with only minor editorial revision. James J. Gehrig, formerly of the staff of the House Committee on Science and Technology, made the last presentation of the final session on "The Rationale for Space Exploration," but his remarks were from notes and are not reproduced here.

Participants brought to the conference a healthy mixture of perspectives from history, political science, journalism, politics, science, and literature. The commentators all were members of the NASA Historical Advisory Committee, which met at New Haven in conjunction with the conference. At the last moment, I.B. Holley, Jr., graciously replaced another committee member who could not attend.

Many hands contributed to the success of the conference and the publication of these proceedings. Special thanks must go to the staff and students of Calhoun and Jonathan Edwards colleges for conceiving the course on which this conference was based and for converting the original idea into a broader undertaking. Both colleges, along with other residential colleges at Yale, also served as hosts to the participants, providing a warm and stimulating atmosphere amidst the rigors of winter in New England. Paul Richenbach and Ann Linbeck were especially helpful and diligent. Monte D. Wright, then Director of the NASA History Office, steered the plan through the bureaucracy with equanimity and skill.

Alex Roland
Durham, North Carolina
1983

SCIENCE, TECHNOLOGY, AND MANAGEMENT:

THE FIRST 20 YEARS IN SPACE

SPACE SCIENCE AND EXPLORATION:
A HISTORICAL PERSPECTIVE*

J.A. Simpson

It is always hazardous to evaluate the historical significance of an era or a development—whether political or intellectual—when the observer is still contemporary with that era. However, when placed against the background of the most significant advances by man throughout history, the space age has a secure position. It is the *evaluation* of the character and significance of the space age, as we shall call it, that we are here to discuss.

Macauley and Livingstone have noted that "many ingredients are necessary for the making of great history . . . knowledge of the facts, truth to record them faithfully, imagination to restore life to dead men and issues. . . . Thucydides had all three ingredients and their union makes him the greatest of historians." I cannot pretend to have these credentials but as a scientist whose main objectives have involved scientific experiments in space and who has shared in some of the space exploration, I can at least present my personal views and perspective. My task is to examine science and exploration in space, not the applications of space science technology. Clearly today the main focus is the U.S. program. But from a historical viewpoint, it is also important to look at the totality of man's efforts in space, in order to recognize the significance of individual achievements within the space era. In this period, six nations (France, Italy, Japan, China, Australia, and the United Kingdom), in addition to the USSR, the European Space Agency, and the United States, have successfully launched their own satellites (app. A). Many other nations have contributed essential experiments or spacecraft for these launchings. My talk here is neither a definitive history or a chronology of developments and achievements in space. It is an overview of the main points of this unique period.

We are all aware of some of the most spectacular and important contributions to our knowledge of the physical world and the universe around us, which have been made by reaching directly to the planets and thereby opening exploration of our solar system. Some of these achievements will be reviewed later. But how does this revolutionary

---

* This paper was supported in part by NASA Grant NGL 14–001–006 and the Arthur H. Compton Fund.

3

step into space compare with other giant strides that have triggered enormous increases in our knowledge and long-term benefits for man? As historical examples we could cite the development of the steam engine and the rise of the industrial revolution, or the achievement of the sustained and controlled nuclear reaction.

In my opinion, some important distinctions should be made among these advances by considering two (and there may be more) kinds of revolutionary developments. The revolutionary development of the *first kind* is one in which a series of critical discoveries were preconditions for the start of the new era or new advance. A recent example is the nuclear age. One can trace the direct steps from James Chadwick's discovery of the neutron (1932) to the Hahn-Meitner discovery of fission of uranium (1939), to the establishment of a sustained nuclear reaction (1942) and, thence, to applications of nuclear energy for both constructive or destructive ends.

I would define a revolution of the *second kind* as the confluence of many ideas and developments, each well known for extended periods of time, which finally come to perfection to trigger revolutionary developments. An example might be Watt's steam engine. His invention of the condenser, to save energy lost in the earlier Newcomen engine, was crucial to the rise of the industrial revolution and represented the revolution's principal technical driver. Concurrent with Watt (1736–1815), Joseph Black evolved the concept of latent heat. This period was followed by Sadi Carnot of France, who was motivated to understand the principles of energy conversion underlying the steam engine by the fact that England had the lead and France was behind in this technology. Even though his ideas were based on an erroneous assumption, he nevertheless laid the groundwork for the basic principles of energy conversion in thermodynamic systems. These examples are intended to show that there are qualitative differences between what I call revolutions of the *first* and *second* kind. The revolution of the *first kind* is a sequential series of discoveries of physical phenomena in nature leading, for example, to a new form of accessible energy. A revolution of the *second kind* has a broad base of many technical developments which, motivated by a need, are finally integrated in a way that leads to further development and a new stage of activity for man.

I believe the achievement of orbiting satellites and probes, as well as manned flight in space and to the Moon (app. B), was a revolution of the *second kind*. Why may we think so? Without recounting the detailed development of rocket power, we know there were two identifiable stages. The first was during World War II when suborbital carriers for destructive weapons were developed; and the second emerged in the 1950s, sparked by the International Geophysical Year (IGY)—a program of scientific exploration and discovery concentrating on the Earth and its surrounding space by scientists in the period 1957–1959. The study of the Earth was not enough. Earth was a part of a larger system involving the space around us that linked phenomena on Earth to the dynamics of the Sun. Consequently, there was a strong consensus among many scientists in the early 1950s that we must go into space with our instrumentation in order to understand the dynamics of the Earth's upper atmosphere, its magnetic field, and related issues. Of course, as recounted in stories throughout the past two centuries, there was always the dream and expectation of someday entering space. But the basis for the strong technological buildup was the need of the scientists, as well as the development of rocket power for national defense. By that time both the United States and the USSR each had the capabilities to launch satellites. Thus, it was only a matter of time until the first satellite, *Sputnik*, was launched successfully by the USSR as part of the International Geophysical Year (IGY) program in science. The success of the USSR effort did not appear to depend on the latest sophisticated technologies. Indeed, while the invention of the transistor in the United States led to the rapid development of electronic technology (which was to become essential for the pursuit of science and exploration in space, and for much of the leadership of U.S. science in space), the Soviet achievement was mainly based on utilizing what was commonly available—what we would call everyday technology of that period. (I can personally verify this since I was invited in 1958 to visit the laboratories where the instrumentation had been built for *Sputnik* and where I could examine firsthand the backup instruments for a *Sputnik*-type spacecraft.) Clearly, in addition to its importance as a political factor, the need to enter space was driven by scientific necessity.

But what are some of the major achievements in space sciences and exploration that could only have come about from activity in space? Before direct entry, the only matter accessible for detailed analysis was mainly from meteorites carrying samples of the early solar system material, and from cosmic rays which are the high-energy nuclei of atoms produced by the nuclear processes associated with the birth and death of stars in the galaxy.

Let us compare our knowledge of specific questions before and after entry into space:

- *Before,* direct entry into space, major questions were open on the nature of the medium between the Sun and the Earth. Was the interplanetary medium, as some believed, virtually a vacuum and static with only occasional interruptions by streams or bursts of particles from the Sun? Or was the medium a dilute gas, perhaps neutral or perhaps partly ionized? It had been deduced that magnetic fields were in interplanetary space. Were these fields continuously present and, if so, how were they distributed through space?

  After, it was proved that there was a continuous flow of ionized gas from the Sun, what we today call the solar wind, rushing outward past the orbit of Earth to the outer boundaries of the solar system. This was one of the alternatives deduced by U.S. experiments and theories prior to 1957, later followed by direct measurements by the USSR and confirmed by U.S. space experiments. The plasma drags a magnetic field, represented by lines of force, outward from the Sun, but since the Sun rotates within an approximately 25-day period, the field lines appear in the form of Archimedes spirals whose pitch depends upon the local speed of the solar wind (see fig. 1).

- *Before,* it was assumed that the Earth's magnetic field extended into space, supporting an equatorial current whose changing characteristics were the source of magnetic storms on Earth, including auroral displays. The only high-energy particles accelerated by natural phenomena known were the cosmic rays, solar flare particles, and auroral particles.

  After, it was found that the Earth's field supported accelerated charged particles and trapped them to form the radiation belts discovered by James Van Allen and confirmed by the USSR.

- *Before,* the general view of the Earth's magnetic field extending into space was dominated by an analogy with an internal source such as a bar magnet (fig. 2), the so-called dipole field.

  After, the Earth's magnetic field was seen as a deformable magnetosphere confined by the solar plasma with the solar wind pressing against the field on the sunward side and dragging the field lines out behind to form a large magnetotail (fig. 3).

- *Before,* the generation of magnetic fields in planets was a controversial subject, and it still is. The radio emission from Jupiter detected from Earth in the 1950s could be explained in terms of a radiation belt around Jupiter two or three times the "size" of the planet, but there was no knowledge concerning the magnetic fields of other planets.

After, Jupiter was found to possess a giant field, full of high-energy particles, extending beyond the solid planet in radius to at least 100 planetary radii (fig. 4). From the *Pioneer* encounter in 1979 and *Voyager* in 1980, Saturn also was found to have a giant magnetic field with characteristics intermediate between Jupiter and Earth. Mercury was a surprise, being found to have a magnetic field and energized particles where none were expected. Mars is still somewhat an enigma with a trivially small field and no evidence of particle acceleration. (The relative sizes of the magnetospheres of the planets is shown in fig. 5.)

- *Before,* the contending views regarding the origin of the Moon extended from assuming that it evolved from the accretion of cold material to assuming that it underwent a heating and mixing cycle similar to that on Earth.

After, the first instruments on the Moon to determine the lunar chemical composition were on the U.S. Surveyor using alpha-particle scattering techniques. The composition showed that the Moon had undergone heating and differentiation (fig. 6) and that the lunar rock was like basalt on Earth. Man's arrival on the Moon was a major technical achievement of the 20th century and samples were brought back which through the radioactive isotopes established the age of the Moon to be about 4 billion years.

- *Before,* planetology based on Earth observations and theory led to conflicting views on Mars, its seasons, and surface features important for deciding on the presence of prehistoric water or cratering by meteorites, etc.

After, the surface features revealed much of the early history of Mars and reduced greatly the probability that some form of life would be found on Mars unless it was prehistoric. The Mars missions stimulated new chemistries, and the dynamics of Mars's atmospheres and polar caps made it possible to understand the seasons on Mars. The Mars missions stimulated renewed experimental interest in defining biophysical definitions of life and life forms and how to test for them.

- *Before,* Mercury appeared only as a fuzzy tennis ball in the highest-powered telescopes.

After, Mercury's surface is heavily cratered, showing that in the early

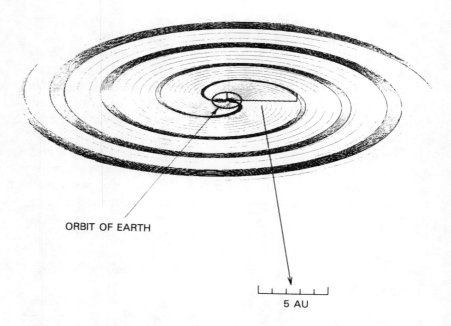

ORBIT OF EARTH

5 AU

Figure 1. Idealized distribution of magnetic field lines of force in interplanetary space near the equatorial plane of the solar system. Magnetic field lines are carried out from the Sun by the solar wind. Spiral-like structure results from the Sun's rotation, which has a period of ∼ 27 days. Concentrations of field lines rooted in solar active centers are regions which sweep past Earth each ∼ 27 days to produce geomagnetic disturbances. (Note: 1 AU is 1 Astronomical Unit, which is the mean distance between Sun and Earth.)

phases of the development of the solar system meteorites were abundant in the inner portion of the solar system, opening a whole new field for planetologists.

• *Before,* the moons of the outer planets were assumed to all have the same origin, although there were various models proposed for the origins of these moons.

After, the Jupiter encounters were the first to reveal that the moons of a planet may be drastically different from each other, as are Callisto or Io. For Saturn the same diversity exists. For example, compare Titan versus Mimas.

• *Before,* Jupiter's atmosphere was an enigma of color bands with four or five spots.

After, we have a startling view of a turbulent atmosphere whose dynamics are only beginning to be understood and which is leading to investigations that will revolutionize our knowledge of planetary atmospheres, including our own atmosphere.

- *Before,* the electromagnetic spectrum used for astronomical observations extended from the radio and infrared to the far ultraviolet.

  After, the useful spectrum was extended to the extreme ultraviolet on through to the x-ray emission from stars and recently to the gamma rays from nuclear processes in our galaxy. Space experiments and observations played an important, and many times crucial, role in the rapid advances in astronomy and astrophysics of the 1950s into the 1980s. They provided much evidence in support of the concept of neutron stars and, later, stars of even higher density—so dense that their gravitational fields prevented light from escaping, the so-called black holes, optically unobservable to an outside observer.

Figure 2. Before the 1950s, Earth's space environment was considered a near-vacuum; the extension of Earth's magnetic field would resemble the field of a simple bar magnet.

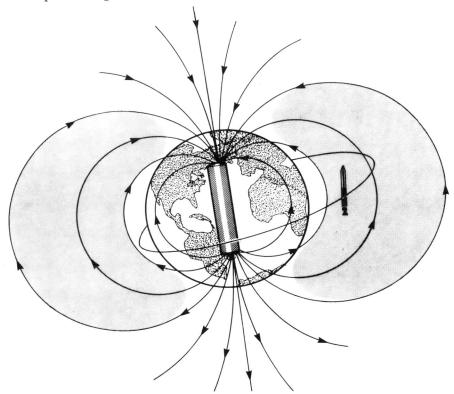

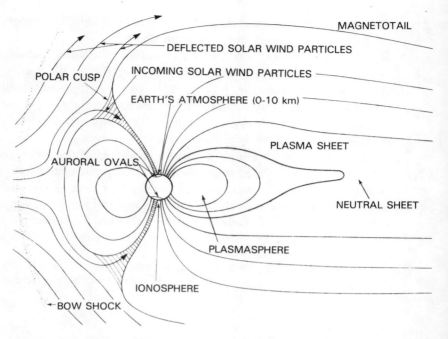

Figure 3. The current concept of geospace (shown here in a noon-midnight meridian plane view) involves a very complex system, and yet even the sophisticated picture is limited by the fact that it has been synthesized from a series of independent measurements collected at different times and places over the past two decades.

The most recent satellite for x-rays is the Einstein Observatory, expanding the regions of universe accessible to us by exploration in the light of the x-rays. These and other observations are providing the quantitative knowledge with which it will become possible to decide whether the universe is closed (and will eventually contract to a singularity), or whether the universe is destined to expand forever.

Even our Sun, viewed in the light of x-rays, reveals totally new aspects of the energetic processes occurring on the surface of the Sun—many of which have a profound impact on conditions on Earth. Furthermore, our view of Earth's atmospheric dynamics is decidedly modified by what has been learned from other planets. On the other hand, it is always difficult, and sometimes impossible, to decide whether or when new essential knowledge on a specific subject would have been acquired even if space vehicles did not exist. This is particularly true in some areas of astrophysics where the continuing development of balloons, high-altitude aircraft, and ground-based in-struments are filling in new areas of the electromagnetic spectrum. An excellent example is ground-based observations of interstellar molecules.

But where do I stop with these examples? Much has been neglected and I must apologize for this sketchy overview.

There are three other novel, but qualitative, aspects of the entry into the space age which belong in our historical perspective.

First, teamwork and government support have combined to yield new approaches to experiments and explorations that are in some ways qualitatively different from the past efforts of a "loner" entrepreneur setting out for exploration. It is now necessary to have "programmed heroes." Only a few can carry out the experiments; only a few per-sonally can enter space, and this rests on competitive processes occur-ring in advance of the event for the selection of scientists, engineers, or

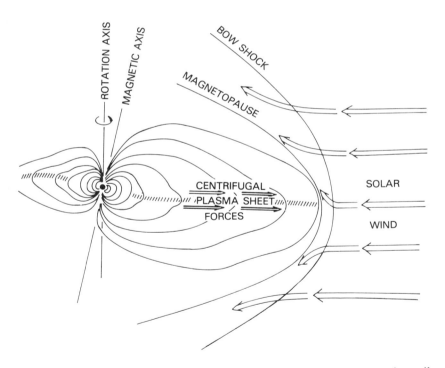

Figure 4. Cross-section sketch of Jupiter and its giant magnetosphere il-lustrating the fact discovered by *Pioneer 10* and *Pioneer 11* that the rotating magnetosphere is an enormous magnetoplasma "machine."

astronauts and their ideas. For the scientist this often means a commitment of a decade or more to obtain approval for a mission and to carry
out an experiment.

Second, there has been and continues to be an extraordinary collaboration among nations for common objectives in space. As examples
I could cite the Apollo-Soyuz or the European Space Agency
(ESA)—National Aeronautics and Space Administration (NASA) International Solar Polar Mission intended to carry spacecraft over the
poles of the solar system in the late 1980s—man's first excursion far
away from the solar equatorial plane (fig. 7).

Perhaps the most significant cooperation, however, is the effort to
establish worldwide treaties for space. An outstanding legacy of the
IGY was the Antarctic Treaty for the scientific exploration of the continent. Hopefully, a legacy of our entry into space will be effective

Figure 5. The relative size of the magnetospheres of the planets is illustrated in
cross-section by assuming that each planet located at the center of the drawing
has the same radius.

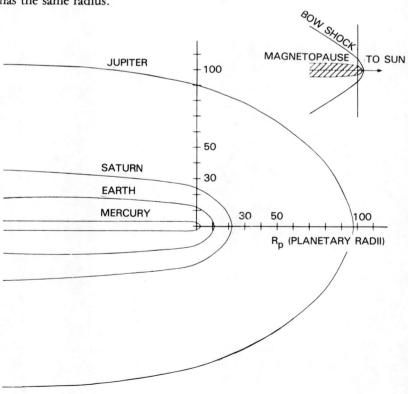

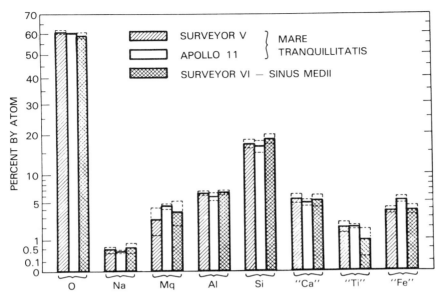

Figure 6. The *Surveyor* spacecraft carrying a University of Chicago experiment weighing ~5 kilograms was first to determine the principal constituents on the Moon which are in close agreement with the later Apollo samples returned to Earth for chemical analysis.

treaties for use and travel in space. The most recent example is the United Nations Moon Treaty (app. C), which is now under review by all nations.

Third, for the first time it has been possible for substantial fractions of the world's population to join the scientists and astronauts in their moments of discovery and exploration, to share in the excitement and wonderment of those moments. This fact, and the pictures of Earth from space, appear to have had an impact on the outlook of millions regarding their place in the universe—a humbling and significant experience for the development of man's concept of himself.

As the most recent example of the participation of the world in discovery, a policy of NASA and the United States, let me cite the encounters of the *Voyager* spacecraft with Saturn which have revealed the fabulous structure of Saturn's rings and atmosphere. These and many more high resolution views were shown on television to the entire world nearly in real time so people throughout the world could participate in the excitement and discovery along with scientists.

For the science and exploration which had been planned in the 1960s and 1970s, we are still succeeding in executing those plans

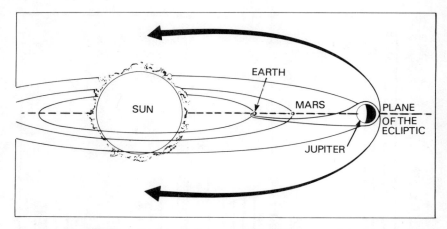

Figure 7. For the first time investigators will be able to send instruments far out of the equatorial plane to obtain three-dimensional studies of phenomena at the Sun, and in interplanetary and interstellar space. The mass of Jupiter will be used as a "slingshot" to enable the spacecraft to travel over the poles of the Sun. (Since this illustration was prepared, the United States has cancelled its spacecraft and only the European spacecraft will be launched in 1986 to be over the poles in ~1989.)

remarkably well. For example, the first generation of space probes (*Pioneer 10* and *11*) are now on their way out of the solar system and may continue to transmit their data to at least 1989–1990 (fig. 8). These probes prove that the United States is invading the solar system. Second-generation *Voyager* spacecraft have now, with sophisticated instruments, followed in the footsteps of *Pioneer 10* and *Pioneer 11*.

The remarkable advances of the USSR—particularly in the areas of early Venus exploration—returned samples from the Moon, and the development of early forms of orbital space stations. Europe, primarily through ESA (app. A-3) is putting its effort heavily on experiments, leaving mainly to the U.S., and soon also to France, the required launch capabilities. Several outstanding examples of European scientific effort include the COS-B, GEO-1, etc. Six other nations are now part of the "club."

Will history show that the United States is now "playing out" the last phases of its leadership in space exploration? It is not at all evident that having taken this lead in space sciences and exploration we in the U.S. will keep it. Even at this stage in our history, there is evidence of uncertainty of commitment by the United States in the face of continued dedication by Europe and the USSR for sustaining a high level

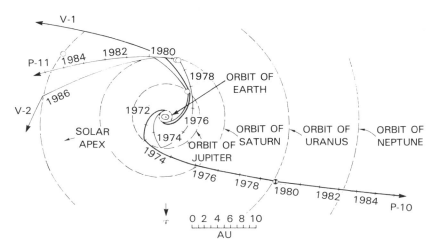

Figure 8. Trajectories of U.S. deep space probes *Pioneer 10* (P-10); *Pioneer 11* (P-11); *Voyager 1* (V-1); and *Voyager 2* (V-2). *Pioneer 10* will transmit data until at least 1990 when it will be beyond the orbits of the planets.

of activity in space. A recent example is the failure of the U.S. to prepare a mission to meet the Halley Comet challenge. There are all too many analogies drawn from history. For example, compare the Spanish explorers and their decline in importance on the seas in the face of Great Britain's major technologies for navigation and naval architecture—a technology on display in Greenwich—which must have played a major role in Britain's dominance of the seas and exploration for centuries.

If nuclear war can be prevented, it appears that we will enter the 21st century greatly troubled over sources of energy, with approximately 80% of the world's population poor, and with dangers of conflict among nations ever present. The many applications derived from space science and exploration and the application of space vehicles to assist in world problems will be crucial in linking the nations of the world.

Finally, it may turn out that the most significant aspect of the entry into space of mankind and his instruments is a new perception people have of their place in the universe, the value of the Earth, and the coming to terms with those factors which could destroy civilization as we know it. This judgment must be left to the historians of the 21st century.

References

Achievements of ESA Scientific Satellites, ESA SP–1013, 1979.

Aeronautics and Space Report of the President: 1980 Activities, (Washington, NASA, 1981).

Frank W. Anderson, Jr. *Orders of Magnitude: A History of NACA and NASA, 1915–1976.* (Washington: NASA, 1976).

Homer E. Newell, *Beyond the Atmosphere: Early Years of Space Science* (Washington: NASA, 1980).

D.E. Page, *Recent Scientific Achievements of ESA Spacecraft.* ESA Bulletin No. 23 (August 1980).

Report Presented by the European Space Agency to the 23rd COSPAR Meeting, Budapest, Hungary, ESA 8–10 Rue Mario-Nikis 75738 (Paris, June 1980).

Thucydides, *The History of the Peloponnesian War,* edited by R. Livingstone (New York: Oxford University Press, 1960).

Appendix A-1. U.S. Spacecraft Record*

(Includes spacecraft from cooperating countries launched by U.S. launch vehicles.)

Year	Earth Orbit[a]		Earth Escape[b]	
	Success	Failure	Success	Failure
1957	0	1	0	0
1958	5	8	0	4
1959	9	9	1	2
1960	16	12	1	2
1961	35	12	0	2
1962	55	12	4	1
1963	62	11	0	0
1964	69	8	4	0
1965	93	7	4	1
1966	94	12	7	1[b]
1967	78	4	10	0
1968	61	15	3	0

Year	Earth Orbit[a]		Earth Escape[b]	
	Success	Failure	Success	Failure
1969	58	1	8	1
1970	36	1	3	0
1971	45	2	8	1
1972	33	2	8	0
1973	23	2	3	0
1974	27	2	1	0
1975	30	4	4	0
1976	33	0	1	0
1977	27	2	2	0
1978	34	2	7	0
1979	18	0	0	0
1980	13	4	0	0
Total	954	133	79	15

[a] The criterion of success or failure used is attainment of Earth orbit or Earth escape rather than judgment of mission success. "Escape" flights include all that were intended to go to at least an altitude equal to lunar distance from the Earth.
[b] This Earth-escape failure did attain Earth orbit and therefore is included in the Earth-orbit success totals.
*From the Aeronautics and Space Report of the President: 1980 Activities, Annual Report of the President to Congress (Washington: NASA, 1981).

Appendix A-2. World Record of Space Launchings Successful in Attaining Earth Orbit or Beyond*

(Enumerates launchings rather than spacecraft; some launches orbited multiple spacecraft.)

Year	United States	USSR	France	Italy	Japan	People's Republic of China	Australia	United Kingdom	European Space Agency	India
1957		2								
1958	5	1								
1959	10	3								
1960	16	3								
1961	29	6								
1962	52	20								
1963	38	17								
1964	57	30								
1965	63	48	1							
1966	75	44	1							
1967	57	66	2	1			1			
1968	45	74								
1969	40	70								
1970	28	81	2	1ª	1	1				
1971	30	83	1	2ª	2	1		1		
1972	30	74		1	1					
1973	23	86								
1974	22	81		2ª	1					
1975	27	89	3	1	2	3				
1976	26	99			1	2				
1977	24	98			2					
1978	32	88			3	1			1	
1979	16	87							1	1
1980	13	89			2					1
Total	756	1339	10	8	17	8	1	1	1	1

ª Includes foreign launchings of U.S. spacecraft.

*From the Aeronautics and Space Report of the President (1980), Annual Report of the President to Congress (Washington: NASA, 1981).

Appendix A-3. ESA/ESRO Scientific Spacecraft Launched

	Launch Date	End of Useful Life	Mission
ESRO-II	May 17, 1968	May 9, 1971	Cosmic rays, solar x-rays
ESRO-IA	October 3, 1968	June 26, 1970	Auroral and polar cap phenomena, ionosphere
HEOS-1	December 5, 1968	October 28, 1975	Interplanetary medium, bow shock
ESRO-1B	October 1, 1969	November 23, 1969	As ESRO-IA
HEOS-2	January 31, 1972	August 2, 1974	Polar magnetosphere, interplanetary medium
TD-1	March 12, 1972	May 4, 1974	Astronomy (UV,x-, and gamma-ray)
ESRO-IV	November 22, 1972	April 15, 1974	Neutral atmosphere, ionosphere, auroral particles
COS-B	August 9, 1975		Gamma-ray astronomy
GEOS-1	April 20, 1977	June 23, 1978	Dynamics of the magnetosphere
ISEE-2	October 22, 1977		Sun/Earth relations and magnetosphere
IUE	January 26, 1978		Ultraviolet astronomy
GEOS-2	July 14, 1978		Magnetospheric fields, waves, and particles

Appendix B. History of U.S. and Soviet Manned Space Flights[*]

Spacecraft	Launch Date	Crew	Flight Time	Highlights
Vostok 1	Apr. 12, 1961	Yuri A. Gagarin	1 h 48 min	First manned flight.
Mercury-Redstone 3	May 5, 1961	Alan B. Shepard, Jr.	15 min	First U.S. flight; suborbital.
Mercury-Redstone 4	July 21, 1961	Virgil I. Grissom	16 min	Suborbital; capsule sank after landing.
Vostok 2	Aug. 6, 1961	Gherman E. Titov	25 h 18 min	First flight exceeding 24 h.
Mercury-Atlas 6	Feb. 20, 1962	John H. Glenn, Jr.	4 hr 55 min	First American to orbit.
Mercury-Atlas 7	May 24, 1962	M. Scott Carpenter	4 h 56 min	Landed 400 km beyond target.
Vostok 3	Aug. 11, 1962	Andrian G. Nikolayev	94 h 22 min	First dual mission (with Vostok 4).
Vostok 4	Aug. 12, 1962	Pavel R. Popovich	70 h 57 min	Came within 6 km of Vostok 3.
Mercury-Atlas 8	Oct. 3, 1962	Walter M. Schirra, Jr.	9 h 13 min	Landed 8 km from target.
Mercury-Atlas 9	May 15, 1963	L. Gordon Cooper, Jr.	34 h 20 min	First U.S. flight exceeding 24 h.
Vostok 5	June 14, 1963	Valeriy F. Bykovskiy	119 h 6 min	Second dual mission (with Vostok 6).
Vostok 6	June 16, 1963	Valentina V. Tereshkova	70 h 50 min	First woman in space: within 5 km of Vostok 5.
Voskhod 1	Oct. 12, 1964	Vladimir M. Komarov Konstantin P. Feoktistov Dr. Boris G. Yegorov	24 h 17 min	First 3-man crew.
Voskhod 2	Mar. 18, 1965	Aleksey A. Leonov Pavel I. Belyayev	26 h 2 min	First extravehicular activity (Leonov, 10 min).
Gemini 3	Mar. 23, 1965	Virgil I. Grissom	4 h 53 min	First U.S. 2-man flight; first manual

[*]From the *Aeronautics and Space Report of the President* (1980), Annual Report of the President to Congress (Washington: NASA, 1981).

Appendix B. (Continued)

Spacecraft	Launch Date	Crew	Flight Time	Highlights
Gemini 4	June 3, 1965	James A. McDivitt Edward H. White, II	97 h 56 min	maneuvers in orbit. 21-min extravehicular activity (White).
Gemini 5	Aug. 21, 1965	L. Gordon Cooper, Jr. Charles Conrad, Jr.	190 h 55 min	Longest-duration manned flight to date.
Gemini 7	Dec. 4, 1965	Frank Borman James A. Lovell, Jr.	350 h 35 min	Longest-duration manned flight to date.
Gemini 6-A	Dec. 15, 1965	Walter M. Schirra, Jr. Thomas P. Stafford	25 h 51 min	Rendezvous within 30 cm of Gemini 7.
Gemini 8	Mar. 16, 1966	Neil A. Armstrong David R. Scott	10 h 41 min	First docking of 2 orbiting spacecraft (Gemini 8 with Agena target rocket).
Gemini 9-A	June 3, 1966	Thomas P. Stafford	72 h 21 min	Extravehicular activity; rendezvous.
Gemini 10	July 18, 1966	John W. Young Michael Collins	70 h 47 min	First dual rendezvous (Gemini 10 with Agena 10, then Agena 8).
Gemini 11	Sept. 12, 1966	Charles Conrad, Jr. Richard F. Gordon, Jr.	71 h 17 min	First initial-orbit docking; first tethered flight; highest Earth-orbit altitude (1,372 km).
Gemini 12	Nov. 11, 1966	James A. Lovell, Jr. Edwin E. Aldrin, Jr.	94 h 35 min	Longest extravehicular activity to date (Aldrin, 5 h 37 min).
Soyuz 1	Apr. 23, 1967	Vladimir M. Komarov	26 hr 37 min	Cosmonaut killed in reentry accident.
Apollo 7	Oct. 11, 1968	Walter M. Schirra, Jr. Donn F. Eisele R. Walter Cunningham	260 h 9 min	First U.S. 3-man mission.

Appendix B. (Continued)

Spacecraft	Launch Date	Crew	Flight Time	Highlights
Soyuz 3 Apollo 8	Oct. 26, 1968 Dec. 21, 1968	Georgiy Beregovoy Frank Borman James A. Lovell, Jr. William A. Anders	94 h 51 min 147 h 1 min	Maneuvered near unmanned Soyuz 2. First manned orbit(s) of Moon; first manned departure from Earth's sphere of influence; highest speed ever attained in manned flight.
Soyuz 4 Soyuz 5	Jan. 14, 1969 Jan. 15, 1969	Vladimir Shatzlov Boris Volynov Aleksey Yeliseyev Yevgeniy Khrunov	71 h 23 min 72 h 56 min	Soyuz 4 and 5 docked and transferred 2 cosmonauts from Soyuz 5 to Soyuz 4.
Apollo 9	Mar. 3, 1969	James A. McDivitt David R. Scott Russell L. Schweickart	241 h 1 min	Successfully simulated in Earth orbit operation of lunar module to landing and take-off from lunar surface and rejoining with command module.
Apollo 10	May 18, 1969	Thomas P. Stafford John W. Young Eugene A. Cernan	192 h 3 min	Successfully demonstrated complete system including lunar module descent to 14,300 m from the lunar surface.
Apollo 11	July 16, 1969	Neil A. Armstrong Michael Collins Edwin E. Aldrin, Jr.	195 h 9 min	First manned landing on lunar surface and safe return to Earth. First return of rock and soil samples to Earth, and manned

Appendix B. (Continued)

Spacecraft	Launch Date	Crew	Flight Time	Highlights
				deployment of experiments on lunar surface.
Soyuz 6	Oct. 11, 1969	Georgiy Shonin Valeriy Kubasov	118 h 42 min	Soyuz 6, 7, and 8 operated as a group flight without actually docking. Each conducted certain experiments, including welding and earth and celestial observation.
Soyuz 7	Oct. 12, 1969	Anatoliy Filipchenko Vladislav Volkov Viktor Gorbatko	118 h 41 min	
Soyuz 8	Oct. 13, 1969	Vladimir Shatalov Aleksey Yeliseyev	118 h 50 min	
Apollo 12	Nov. 14, 1969	Charles Conrad, Jr. Richard F. Gordon, Jr. Alan L. Bean	244 h 36 min	Second manned lunar landing. Continued manned exploration and retrieved parts of Surveyor III spacecraft which landed in Ocean of Storms on Apr. 19, 1967.
Apollo 13	Apr. 11, 1970	James A. Lovell, Jr. Fred W. Haise, Jr. John L. Swigert, Jr.	142 h 55 min	Mission aborted due to explosion in the service module. Ship circled Moon, with crew using LEM as "lifeboat" until just prior to reentry.
Soyuz 9	June 1, 1970	Andrian G. Nikolayev Vitaliy I. Sevastianov	424 h 59 min	Longest manned spaceflight to date, lasting 17 days 16h 59 min.
Apollo 14	Jan. 31, 1971	Alan B. Shephard, Jr. Stuart A. Roosa Edgar D. Mitchell	216 h 2 min	Third manned lunar landing. Mission demonstrated pinpoint landing capability and continued manned exploration.
Soyuz 10	Apr. 22, 1971	Vladimir Shatalov	47 h 48 min	Docked with Salyut 1, but crew did not

Appendix B. (Continued)

Spacecraft	Launch Date	Crew	Flight Time	Highlights
		Aleksey Yeliseyev Nikolai Rukavishnikov		board space station launched Apr. 19. Crew recovered Apr. 24, 1971.
Soyuz 11	June 6, 1971	Georgiy Timofreyevich Dobrovolskiy Vladislav Nikolayevich Volkov Viktor Ivanovich Patsayev	570 h 22 min	Docked with Salyut 1 and Soyuz 11 crew occupied space station for 22 days. Crew perished during final phase of Soyuz 11 capsule recovery on June 30, 1971.
Apollo 15	July 26, 1971	David R. Scott Alfred M. Worden James Bensen Irwin	295 h 12 min	Fourth manned lunar landing and first Apollo "J" series mission which carry the Lunar Roving Vehicle. Worden's in-flight EVA of 38 min 12 s was performed during return trip.
Apollo 16	Apr. 16, 1972	John W. Young Charles M. Duke, Jr. Thomas K. Mattingly, II	265 h 51 min	Fifth manned lunar landing, with Lunar Roving Vehicle.
Apollo 17	Dec. 7, 1972	Eugene A. Cernan Harrison H. Schmitt Ronald E. Evans	301 h 52 min	Sixth and final Apollo manned lunar landing, again with roving vehicle.
Skylab 2	May 25, 1973	Charles Conrad, Jr. Joseph P. Kerwin	627 h 50 min	Docked with Skylab 1 for 28 days. Repaired damaged station.

Appendix B. (Continued)

Spacecraft	Launch Date	Crew	Flight Time	Highlights
Skylab 3	July 28, 1973	Paul J. Weitz Alan L. Bean Jack R. Lousma Owen K. Garriott	1427 h 9 min	Docked with Skylab 1 for over 59 days.
Soyuz 12	Sept. 27, 1973	Vasiliy Lazarev Oleg Makarov	47 h 16 min	Checkout of improved Soyuz.
Skylab 4	Nov. 16, 1973	Gerald P. Carr Edward G. Gibson William R. Pogue	2017 h 16 min	Docked with Skylab 1 in long-duration mission; last of Skylab program.
Soyuz 13	Dec. 18, 1973	Petr Klimuk Valentin Lebedev	188 h 55 min	Astrophysical, biological, and Earth resources experiments.
Soyuz 14	July 3, 1974	Pavel Popovich Yuriy Artyukhin	377 h 30 min	Docked with Salyut 3 and Soyuz 14 crew occupied space station for over 14 days.
Soyuz 15	Aug. 26, 1974	Gennadiy Sarafanov Lev Demin	48 h 12 min	Rendezvoused but did not dock with Salyut 3.
Soyuz 16	Dec. 2, 1974	Anatoliy Filipchenko Nikolai Rukavishnikov	142 h 24 min	Test of ASTP configuration.
Soyuz 17	Jan. 10, 1975	Aleksey Gubarev Georgiy Grechko	709 h 20 min	Docked with Salyut 4 and occupied station during a 29-day flight.
Anomaly	Apr. 5, 1975	Vasiley Lazarev Oleg Makarov	20 min	Soyuz stages failed to separate; crew recovered after abort.

Appendix B. (Continued)

Spacecraft	Launch Date	Crew	Flight Time	Highlights
Soyuz 18	May 24, 1975	Petr Klimuk Vitaliy Sevastiyanov	1,511 h 20 min	Docked with Salyut 4 and occupied station during a 63-day mission.
Soyuz 19	July 15, 1975	Aleksey Leonov Valeriy Kubasov	142 h 31 min	Target for Apollo in docking and joint experiments ASTP mission.
Apollo	July 15, 1975	Thomas P. Stafford Donald K. Slayton Vance D. Brand	217 h 28 min	Docked with Soyuz 19 in joint experiments of ASTP mission.
Soyuz 21	July 6, 1976	Boris Volynov Vitaliy Zholobov	1,182 h 24 min	Docked with Salyut 5 and occupied station during 49-day flight.
Soyuz 22	Sept. 15, 1976	Valeriy Bykovskiy Vladimir Aksenov	189 h 54 min	Earth resources study with multispectral camera system.
Soyuz 23	Oct. 14, 1976	Vyacheslav Zudov Valeriy Rozhdestvenskiy	48 h 6 min	Failed to dock with Salyut 5.
Soyuz 24	Feb. 7, 1977	Viktor Gorbatko Yuriy Glazkov	425 h 23 min	Docked with Salyut 5 and occupied station during 18-day flight.
Soyuz 25	Oct. 9, 1977	Vladimir Kovalenok Valeriy Ryumin	48 h 46 min	Failed to achieve hard dock with Salyut 6 station.
Soyuz 26	Dec. 10, 1977	Yuriy V. Romanenko Georgiy M. Grechko	898 h 6 min	Docked with Salyut 6. Crew returned in Soyuz 27; crew duration 2,314 h.
Soyuz 27	Jan. 10, 1978	Vladimir A. Dzhanibekov Oleg G. Makarov	1558 h 53 min	Docked with Salyut 6. Crew returned in Soyuz 26; crew duration 142 h 59 min.

Appendix B. (Continued)

Spacecraft	Launch Date	Crew	Flight Time	Highlights
Soyuz 28	Mar. 2, 1978	Aleksey A. Gubarev Vladimir Remek	190 h 17 min	Docked with Salyut 6. Remek was first Czech cosmonaut to orbit.
Soyuz 29	June 15, 1978	Vladimir V. Kovalenok Aleksandr S. Ivanchenkov	1,911 h 23 min	Docked with Salyut 6. Crew returned in Soyuz 31; crew duration 3,350 h 48 min.
Soyuz 30	June 27, 1978	Petr I. Klimuk Miroslaw Hermaszrewski	190 h 4 min	Docked with Salyut 6. Hermaszewski was first Polish cosmonaut to orbit.
Soyuz 31	Aug. 26, 1978	Valeriy F. Bykovskiy Sigmund Jaehn	1,628 h 14 min	Docked with Salyut 6. Crew returned in Soyuz 29; crew duration 188 h 49 min. Jaehn was first German Democratic Republic cosmonaut to orbit.
Soyuz 32	Feb. 25, 1979	Vladimir A. Lyakhov Valeriy V. Ryumin	2,596 h 24 min	Docked with Salyut 6. Crew returned in Soyuz 34; crew duration 4200 h 36 min, or 175 days.
Soyuz 33	Apr. 10, 1979	Nikolay N. Rukavishnikov Georgi I. Ivanov	47 h 1 min	Failed to achieve docking with Salyut 6 station. Ivanov was first Bulgarian cosmonaut to orbit.
Soyuz 34	June 6, 1979	(unmanned at launch)	1,770 h 17 min	Docked with Salyut 6, later served as ferry for Soyuz 32 crew while Soyuz 32 returned unmanned.
Soyuz 35	Apr. 9, 1980	Leonid I. Popov Valeriy V. Ryumin	1,321 h 29 min	Docked with Salyut 6. Crew returned in Soyuz 37; crew duration 4,436 h 12 min.

Appendix B. (Continued)

Spacecraft	Launch Date	Crew	Flight Time	Highlights
Soyuz 36	May 26, 1980	Valeriy N. Kubasov Bertalan Farkas	1,580 h 54 min	Docked with Salyut 6. Crew returned in Soyuz 35; crew duration 188 h 46 min. Farkas was first Hungarian to orbit.
Soyuz T-2	June 5, 1980	Yuriy V. Malyshev Vladimir V. Aksenov	94 h 21 min	Docked with Salyut 6. First manned flight of new generation ferry.
Soyuz 37	July 23, 1980	Viktor V. Gorbatko Pham Tuan	1,911 h 17 min	Docked with Salyut 6. Crew returned in Soyuz 36; crew duration 188 h 42 min. Pham was first Vietnamese to orbit.
Soyuz 38	Sept. 18, 1980	Yuriy V. Romanenko Arnaldo Tamayo Mendez	188 h 43 min	Docked with Salyut 6. Tamayo was first Cuban to orbit.
Soyuz T-3	Nov. 27, 1980	Leonid D. Kizim Oleg G. Makarov Gennadiy M. Strekalov	307 h 8 min	Docked with Salyut 6. First 3-man flight in Soviet program since 1971.

Appendix C. The United Nations Moon Treaty

The Moon Treaty has been under discussion since late 1971 when the General Assembly adopted resolution 2779, in which it took note of a draft treaty submitted by the USSR and requested the Committee on the Peaceful Uses of Outer Space (COPUOS) and its legal Subcommittee (LSC) to consider the question of the elaboration of a draft international treaty concerning the Moon on a priority basis.

The draft Moon Treaty is based to a considerable extent on the 1967 Outer Space Treaty. Indeed, the discussion in the Outer Space Committee confirmed the understanding that the Moon Treaty in no way derogates from or limits the provisions of the 1967 Outer Space Treaty.

The draft Moon Treaty also is, in its own right, a meaningful advance in the codification of international law dealing with outer space, containing obligations of both immediate and long-term application to such matters as the safeguarding of human life on celestial bodies, the promotion of scientific investigation and the exchange of information relative to and derived from activities on celestial bodies, and the enhancement of opportunities and conditions for evaluation, research, and exploitation of the natural resources of celestial bodies.

The General Assembly, by consensus, opened the treaty for signature on December 5, 1979.

This appendix presents the text of the draft treaty in the left column on each page; in the right column, opposite the appropriate sections of the text, are some comments by the Department of State on the attitude of the United States regarding particular provisions.

Treaty Text	Commentary by Department of State

Draft agreement governing the activities of States on the moon and other celestial bodies.

The States Parties to this Agreement,

Noting the achievements of States in the exploration and use of the moon and other celestial bodies,

Recognizing that the moon, as a natural satellite of the earth, has an important role to play in the exploration of outer space,

Determined to promote on the basis of equality the further development of co-operation among States in the exploration and use of the moon and other celestial bodies,

Desiring to prevent the moon from becoming an area of international conflict,

Bearing in mind the benefits which may be derived from the exploitation of the natural resources of the moon and other celestial bodies,

Recalling the Treaty on Principles Governing the Activities of States in the Exploration and Use of Outer Space, including the Moon and Other Celestial Bodies, the Agreement on the Rescue of Astronauts, the Return of Astronauts and the Return of Objects Launched into Outer Space, the Convention on International Liability for Damage Caused by Space Objects, and the Convention on Registration of Objects Launched into Outer Space.

Taking into account the need to define and develop the provisions of these international instruments in relation to the moon and other celestial bodies, having regard to further progress in the exploration and use of outer space,

Have agreed on the following:

Article I

1. The provisions of this Agreement relating to the moon shall also apply to other celestial bodies within the solar system, other than the earth, except in so far as specific legal norms enter into force with respect to any of these celestial bodies.

2. For the purposes of this Agreement reference to the moon shall include orbits around or other trajectories to or around it.

3. This Agreement does not apply to extraterrestrial materials which reach the surface of the earth by natural means.

There has been considerable discussion of Article I of the draft treaty. The United States accepts the Outer Space Committee's conclusions as to this article—namely, first, that references to the moon are intended also to the references to other celestial bodies within our solar system other than the earth; secondly, that references to the moon's natural resources are intended to comprehend those natural resources to be found on these celestial bodies; and, thirdly that the trajectories and orbits referred to in Article I, paragraph 2, do not include trajectories and orbits of space objects between the earth and earth orbit or in earth orbit only. In regard to the phrase "earth orbit only", the fact that a space object in earth orbit also is in orbit around the sun does not bring space objects which are only in earth orbit within the scope of this treaty.

Treaty Text	**Commentary by Department of State**

Article II

All activities on the moon, including its exploration and use, shall be carried out in accordance with international law, in particular the Charter of the United Nations, and taking into account the Declaration on Principles of International Law concerning Friendly Relations and Cooperation among States in accordance with the Charter of the United Nations, adopted by the General Assembly on 24 October 1970, in the interest of maintaining international peace and security and promoting international co-operation and mutual understanding, and with due regard to the corresponding interests of all other States Parties.

Article II reaffirms the application of the Charter of the United Nations and of international law to outer space. While the Charter predates man's entry into space, its principles and provisions, including those relating to the permissible and impermissible uses of force, are as valid for outer space as they are for our seas, land, or air. The United States welcomes the international community's reaffirmation in the Moon Treaty of this essential point.

Article III

1. The moon shall be used by all States Parties exclusively for peaceful purposes.

2. Any threat or use of force or any other hostile act on the moon is prohibited. It is likewise prohibited to use the moon in order to commit any such act or to engage in any such threat in relation to the earth, the moon, spacecraft, the personnel of spacecraft or manmade objects.

Article III contains a statement of the principle that the celestial bodies and those orbits around them and to them are only to be used for peaceful—i.e., nonaggressive—purposes.

Paragraph 2 of Article III spells out in some detail some of the consequences to be drawn from Article II. Specifically, paragraph 2's purpose is to make clear that it is forbidden for a party to the Moon Treaty to engage in any threat or use of force on the moon or in other circumstances set forth in paragraph 2 if such acts would constitute a violation of the party's international obligations in regard to the threat or use of force.

3. States Parties shall not place in orbit around or other trajectory to or around the moon objects carrying nuclear weapons or any other kinds of weapons of mass destruction or place or use such weapons on or in the moon.

4. The establishment of military bases, installations and fortifications, the testing of any type of weapons and the conduct of military manoeuvres on the moon shall be forbidden. The use of military personnel for scientific research or for any other peaceful purposes shall not be prohibited. The use of any equipment or facility necessary for peaceful exploration and use of the moon shall also not be prohibited.

Article IV

1. The exploration and use of the moon shall be the province of all mankind and shall be carried out for the benefit and in the interests of all countries, irrespective of their degree of economic or scientific development. Due regard shall be paid to the interest of present and future generations as well as to the need to promote higher standards of living conditions of economic and social progress and development in accordance with the Charter of the United Nations.

SATELLITES AND POLITICS:
WEATHER, COMMUNICATIONS, AND EARTH RESOURCES

Pamela Mack

Since its founding in 1958, the National Aeronautics and Space Administration (NASA) has concentrated its effort in developing practical
uses for spaceflight, or space applications, in three programs: weather,
communications, and Earth resources satellites. Weather satellites and
communications satellites have been tested and improved so that they
have now reached the stage of routine or operational use, but Earth
resources satellites are still experimental.

With applications satellites, NASA had to solve an extra problem
not present in most other space projects: these satellites were developed
for users outside of NASA. W. Henry Lambright, among others, has
pointed out that conflict often arises when the agency developing a new
technology is not responsible to the agency that will actually use it. The
history of the three applications satellite programs shows different kinds
of problems that can arise from this situation depending on the relative
power of the various players, the divergence of their interests, and uses to
which the satellites can be put.

For weather satellites, problems between NASA and the user agency
arose only when the program was nearly ready to make the transition to an
operational system. This was true not because of effective cooperation
with the user, the Weather Bureau, but because of lack of coordination.

Weather satellites use a television-type camera to take pictures of
cloud cover and then radio the pictures to Earth. Two types of weather
satellites are now used: low altitude satellites, which rapidly orbit the
Earth taking pictures of various areas, and geosynchronous satellites,
which orbit at such an altitude that they always remain over the same
point of the Earth's surface and therefore provide continuous monitoring
of the weather on one half of the globe. Communication technology has
been improved so that the satellites now continuously broadcast the
television pictures they take. These pictures can be received and used by
anyone with an inexpensive antenna and printer. The first weather
satellites proved immediately useful for tracking hurricanes and other
large-scale features difficult to observe as a whole from the ground. The
benefits to routine weather forecasting have been limited, however, by
the lack of a model of the atmosphere exact enough to provide completely
accurate predictions even from plentiful data.

Research on the possibility of using satellites to monitor weather started as a military project. The project was transferred to NASA in 1959, under President Eisenhower's commitment to put as much of the space program as possible in civilian hands. The Weather Bureau had little voice in NASA's program; NASA formed an interagency advisory committee, but it had little influence. When the first weather satellite, Tiros, was launched in 1960, NASA asked the Weather Bureau to analyze the data. Meteorologists found the data very useful, and within a few days the Weather Bureau started making cloud-cover maps from satellite data and distributing them to meteorologists to aid in making routine forecasts.

NASA planned to follow the experimental Tiros project with a more sophisticated series of proto-operational satellites called Nimbus. The Weather Bureau, however, found the Tiros data satisfactory and was suspicious of the plans for Nimbus because it was very expensive and might not be ready before the last Tiros satellite reached the end of its useful life. The Weather Bureau did not want to commit itself to an expensive satellite program which, once operational, would be paid for entirely from the Bureau's small budget. On September 27, 1963, the Weather Bureau officially notified NASA that it was withdrawing from the Nimbus programs and the existing interagency agreement, and proposed an interim operational satellite based on Tiros and a new agreement making NASA and the Weather Bureau equal partners. The Weather Bureau, a weak agency without much support from its parent institution, the Department of Commerce, could afford to make such a move only because it had found a backer. The Department of Defense offered to cooperate with the Weather Bureau and provide the necessary expertise with space hardware if NASA refused to meet the Weather Bureau's terms. Defense was jealous of NASA for taking over projects from the military space program and was concerned about the possibility of a gap between the Tiros and Nimbus programs that would leave the military without storm-warning information it already depended on. Faced with losing the whole program, NASA negotiated a new agreement with the Weather Bureau for a Tiros operational system.

In this case the political conflict grew out of the divergence of interests of the research agency and the user agency. NASA wanted to develop a second generation of satellites employing the most sophisticated technology, while the Weather Bureau wanted to use the simpler, less expensive system already in hand and not yet fully utilized. The Weather Bureau wanted one sort of satellite and NASA wanted another, but instead of compromising, NASA simply ignored the Weather Bureau. This naturally resulted in trouble when the time came for the Weather Bureau to start planning to take over the system from NASA. The location of the research function in the operating agency, the Weather Bureau, would have slowed down the advance of new technology, but perhaps learning to use the old technology better would have been (and was) more productive. Research groups tend towards independence, whether they are separate or located in operating agencies, and researchers can rarely see that more sophisticated technology is not necessarily more useful.

In the case of communications satellites, the problem of transition from an experimental to an operational system was compounded by conflict over who would be the operational user. The communications industry saw the possibility of large profits, and the Congress had to deal with tricky philosophical issues of public versus private control.

Communications satellites relay radio waves carrying telephone, television, and data signals from one point on Earth to another. NASA tested three varieties. Passive satellites, like Echo, simply provide a reflective surface for radio waves to bounce off. Echo is just a giant mylar balloon. Active satellites, which come in two types, receive the signal from the ground, amplify it, and retransmit it to its destination. Low altitude active satellites, like Relay and Telstar, move rapidly relative to the surface of the Earth. This means that the antenna on the ground must be pointed to follow the satellite and a number of satellites are needed so that one is always available above the horizon. Geosynchronous active satellites, like Syncom, are placed in such an orbit that they remain always over the same point on the Earth's surface. This more distant orbit requires more powerful transmitters and more sensitive receivers on the satellite and the ground, but the advantages of the fixed position are more important. Almost all of the many operational communications satellites currently in use are of this type.

NASA started out with a limited role in communications satellite research—first only passive satellites, then only low-altitude satellites—because of a division of responsibilities with the Department of Defense. Unlike other applications programs, however, this type of satellite was clearly going to be profitable to private industry, which therefore set the pace. American Telephone and Telegraph (AT&T) and, on a smaller scale, other companies spent their own funds on communications satellite research in hopes of getting lucrative contracts later, or, in the case of AT&T, in hopes of gaining a monopoly. AT&T developed its own low-altitude, active, experimental satellite, Telstar, and requested that NASA launch it. This would have put AT&T in a strong position to launch the first communication satellite system as a private venture.

Because of concerns about monopoly, diplomacy, and giving away the fruits of government research, private industry did not get the free rein it wanted. NASA insisted that a government-funded and government-controlled experimental communications satellite, to be developed under a contract awarded by competitive bidding (to Hughes Aircraft Co.), be planned first. NASA envisioned that after its experimental program, Relay, an operational communications satellite system would be owned by private industry. NASA launched AT&T's satellite in July 1962 after awarding the contract for Relay, but before its launch. Meanwhile, the Congress fought over details of the institutional arrangements for the operational system. The Department of State was concerned over a private company controlling the U.S. share of an international communications system; liberals did not want to see government research given away for private profit; conservatives wanted the government out of a function that private industry could handle; and communications and aerospace firms wanted as much of the control and profits as possible. The end result was COMSAT, a private company with some board members appointed by the President, carefully defined federal jurisdictions, and broad ownership by communications and aerospace firms and the general public.

This political fight slowed the development of the technology and altered its character. During the political controversy, NASA proceeded with research on a geosynchronous communications satellite, too advanced for the private companies to risk on their own. The tests of this satellite, Syncoms I and II, launched in February and July 1963, proved very successful. For the first operational communications satellite system, COMSAT chose to develop not the system of low altitude satellites that AT&T and the other communications companies had planned on, but rather a much less expensive system of geosynchronous satellites. In this case, unlike that of meteorological satellites, the users were grateful for the advanced technology that NASA had developed despite their initial lack of interest.

The transition from an experimental to an operational system of communications satellites was disrupted by disagreements more over political philosophy than over technology. The technology was affected, however, when the political arguments provided extra time during which a new technology proved to be superior. AT&T had wanted to gain control over the system by being the first to develop the technology. The company failed to get economic control or contracts for its technology as a whole, but the effort no doubt strengthened its position in Comsat and the component market.

For Earth resources satellites, NASA had to deal with a wide variety of users, leaving the goals of the program uncertain. Without a clear idea of who would use the satellite for what, choices of technology were controversial.

Earth resources satellites provide wide-scale, repetitive pictures of the surface of the Earth for the survey and monitoring of resources. The first *Landsat* satellite was launched in 1972; the second and third are still functioning and carry two sensors: a kind of television camera and a scanner that provides more precise color data. The satellite radios the data to Earth, where it is printed on photographic film or analyzed by a computer. Even at the present coarse resolution of 60 to 100 meters, the satellite radios down 15 million bits of data per second. Processing, storing, and extracting information from this flood of data have proved to be the most difficult technological challenge of the project. The data have been used successfully, at least on an experimental scale, to detect large geological features associated with oil and minerals, to measure the areas planted in different crops (to help predict harvests), to monitor water distribution and snow cover to predict flooding, and to make maps of land use. Users include federal, state, and local government agencies and private firms.

The federal agencies were the only users with a voice in the development of the first satellite. NASA set up a program in 1964 to investigate the use of space vehicles to study Earth resources and transferred money to the departments of the Interior and Agriculture to consider what use they could make of the data. The Department of the Interior developed so much enthusiasm for the idea that when NASA moved slowly in making plans for an experimental satellite, Interior pushed the project along by announcing its own satellite program. An independent satellite project was vetoed by the President because experimental satellites were NASA's domain, but NASA speeded up its project. The Department of Agriculture proposed a different sensor from that desired by Interior. Each agency pushed for a small, simple satellite with the sensor that would make the satellite most useful to the agency. NASA compromised by flying both sensors and choosing spectral bands useful for the widest possible range of applications. Some users have complained that these spectral bands make the data difficult to use because they are not optimal for any application. Compromises were also made in the choice of orbit and NASA settled for two sensors instead of the more elaborate experiment it had originally proposed.

To further complicate the situation, NASA soon realized that some of the greatest benefits from *Landsat* would come from improved resource management on the state and local level. NASA had developed the satellite without consulting these users, and it proved difficult to persuade them to use the new information. NASA set up a technology transfer program for *Landsat*, which started out just publicizing information but has gradually developed joint projects that are effective in convincing states to use *Landsat* data. The states have been reluctant to participate because of distrust of sophisticated technology, which NASA as an agency seems to symbolize, and because they did not want to make an investment until the program had settled into a final operational form. Because of the lack of immediate benefits and wide use after the 1972 launch of the first satellite, the Office of Management and Budget has opposed the transition of *Landsat* from an experimental project into an operational program. The commitment to an operational program, to be managed by the National Oceanic and Atmospheric Administration, was made only in late 1979.

In the case of *Landsat*, NASA successfully played the users off against each other so that none had control, but the result was a project with a shortage of goals and support. The users NASA was most interested in, state and local governments, had not asked for the project or shaped the system into something useful to them. Because of this and their lack of technological sophistication, they had little interest in adopting the new techniques NASA had developed. Perhaps with more involvement of the users in the design and more understanding of the diffusion of new techniques, the project would have brought more benefits by now. In any case, the politics of balancing the demanding agency users and the concept of future state and local users forced NASA to choose the most neutral technology—useful to everyone but ideal for no use. NASA provided different technology than individual users wanted in order to make one satellite serve the whole range of users. The combination satellite is not completely satisfactory, but the Office of Management and Budget would probably not have approved more than one satellite.

NASA has found the process of developing satellite programs for other agencies fraught with controversy. The space agency has, probably unavoidably, looked after its own interests in expanding its research program and pursued advancing technology without much sensitivity to the needs of the eventual users. The problem is a tricky one, however, because NASA can claim with some validity that the users, because they are not technologically sophisticated, do not realize the potential benefits of new technology. The three cases of applications satellites show the users as sometimes grateful and sometimes not for the technology developed despite their wishes. The answer, I believe, lies not in a better balance between the users' demands and NASA's ideas, but in taking the trouble to educate the users to participate in the development of the technology.

Source Notes

My information on weather satellites comes from Richard LeRoy Chapman's excellent dissertation (Syracuse University, 1967) *A Case Study of the U.S. Weather Satellite Program: The Interaction of Science and Politics*. Chapman, like W. Henry Lambridge in *Governing Science and Technology* (New York: Oxford University Press, 1976), emphasizes the transition from an experimental to an operational system as a key policy problem.

My discussion of communications satellites is based mostly on Jonathan F. Galloway, *The Politics and Technology of Satellite Communications* (Lexington, Mass.: Lexington Books of D.C. Heath and Co., 1972), and Delbert D. Smith, *Communication Via Satellite: A Vision in Retrospect* (Leyden, Boston: A.W. Sijthott, 1976). I also looked at Michael E. Kinsley, *Outer Space and Inner Sanctums: Government, Business, and Satellite Communications* (New York: John Wiley & Sons, 1976), a Nader report; J.R. Pierce, *The Beginning of Satellite Communications* (San Francisco: San Francisco Press, 1968), giving the AT&T view; and Roger A. Kvam, "Comsat: The Inevitable Anomaly," in Stanford A. Lakoff, ed., *Knowledge and Power: Essays on Science and Government* (New York: The Free Press, 1966).

There are no useful secondary sources on the history of Earth resources satellites except for W. Henry Lambright, "ERTS: Notes on a 'Leisurely' Technology," *Public Science Newsletter* (Aug.-Sept. 1973), pp. 1–8. The information presented here is based on archival research at NASA and the Department of the Interior for my dissertation, "The Politics of Technological Change: A History of Landsat" (University of Pennsylvania, 1984).

I would like to thank Alex Roland and John Mack for criticism and comments and the NASA History Office and the National Air and Space Museum for financial support of the dissertation.

MANAGEMENT OF LARGE-SCALE TECHNOLOGY

Arnold S. Levine

The history of the United States space program in the 1960s has the appeal of something conceived with magnificent simplicity and carried out on the grand scale. Between 1961 and 1970, the National Aeronautics and Space Administration (NASA) launched several dozen unmanned spacecraft, revolutionizing communications and meteorological technologies, on the one hand, and electronics and software development on the other. But in the public mind, NASA was most closely associated with the manned spaceflight programs—Project Mercury (1958-1963), which tested the ability of one man to function up to several hours in Earth orbit; Gemini (1962-1966), in which two men in one spacecraft were assigned a variety of tasks, including rendezvous and docking in Earth orbit with a target vehicle and moving around outside the spacecraft itself; and Apollo (1961-1972), wherein three-man crews were sent on progressively more ambitious missions, culminating in the lunar landing of July 1969. Merely to sketch the civilian space program thus is to indicate the magnitude of NASA assignments and the scope of its successes. One must take seriously the contention of James E. Webb, NASA Administrator from 1961 to 1968, that the success of NASA was a success in organizing "large-scale endeavors," i.e., that the same system of management that made the lunar landings possible may also have been their most important byproduct.

In this paper, I am going to try to answer the following question: What can the study of NASA, as an organization, teach us? Using Webb's concept of the large-scale endeavor as a starting point, I will concentrate on NASA as a going concern; in other words, as an organization that, instituted for specific purposes, strove to maintain itself, to operate within the terms of its establishment, and to compete with other agencies for the limited resources made available by Congress and the White House. Put differently, themes running through this paper will be: (1) how a high-technology agency was run in a decade marked by rapid expansion of funds and manpower in the first half and almost as rapid contraction in the second; and (2) how NASA combined centralized planning and control with decentralized project execution. In turn, each of these themes raises subsidiary questions: What criteria did the agency use in choosing its contractors and, in the absence of market conditions, how did it supervise them to get the hardware and services for which it contracted? How did NASA maintain its independence vis-à-vis the Department of

Defense (DoD), the one federal agency with which NASA had to come to terms?

The concept of the large-scale endeavor is useful but, at the same time, difficult to pin down. In his *Space Age Management,* drafted in his last months at NASA and published shortly after he resigned as administrator, Webb discussed the characteristics of the large-scale endeavor. Typically, the endeavor results from a new and urgent need or a new opportunity created by social, political, technological, or military changes in the environment. Most often, it requires "doing something for the first time and [has] a high degree of uncertainty as to precise results," and it will have second- and third-order consequences, often unintended, beyond the main objective.[1] Finally, such endeavors "do not generally require new organizational and administrative forms, but the more effective utilization of existing forms."[2] Webb's description can, of course, apply to many endeavors beside the space program; the attempt to build and operate a national rail passenger network, to develop a strategic petroleum reserve, to build the Alaska pipeline, or to conduct the War on Poverty—all share many of the features Webb enumerates. But the space program and the projects comprising it had certain advantages in attaining its goals, stemming from the nature of its mission, which most of the endeavors named above lacked.

First, the NASA goals could be stated in precise, operational terms. The agency would describe a goal within the broader mission: put a communications satellite in synchronous Earth orbit; or, develop an unmanned spacecraft to soft-land on the Moon and a vehicle with a liquid-hydrogen upper stage to launch it. Such precision may be contrasted with those federal agencies charged with improving the quality of education, fighting alcoholism and drug abuse, or finding permanent jobs for the hard-core unemployed. As Charles Lindblom and David Cohen have noted, "Government agencies are again and again assigned . . . responsibilities beyond any person's or organization's known competence. They do not typically resist these assignments because they are funded and maintained for their efforts, not for their results."[3]

Second, NASA in the early 1960s had an organizational flexibility unmatched by any agency of comparable size. In this period NASA had no formal agency-wide long-range plan; no general advisory committee of outside scientists, such as those established for the Atomic Energy Commission and the Department of Defense; no inspector-general, chief scientist, or chief engineer; no centralized range structure for tracking, data acquisition, and mission control; no central planning staff attached to the Office of the Administrator. These functions were handled in other, much more decentralized ways. Moreover, the absence of a plan or general advisory committee rescued the agency from becoming captive to policies which might cease to be relevant. To maintain this flexibility and to adapt the agency to change, there were frequent reorganizations, notably in 1961, 1963, 1965, and 1967. But they were not ends in themselves. They were designed less to set certain things right—for instance, to improve communications between decision-makers and their supporting staffs, or to free the field centers from unneeded supervision—than to turn the agency from one set of programs to those of quite a different sort. For NASA was vulnerable. It had to stake a claim to territory of its own, rather than becoming (as its predecessor, the National Advisory Committee for Aeronautics, had been) a supporting arm of the military services, or a supervisory agency with a small in-house staff and contractor-operated facilities, like the Atomic Energy Commission.

Finally, NASA in the 1960s was an agency with a single mission—to land a man on the Moon and return him safely before the end of the decade—but with numerous subordinate goals. The National Aeronautics and Space Act enacted by Congress in July 1958 was permissive rather than mandatory, so far as ends were concerned. It was a shopping list as much as an enabling act, freeing NASA to pursue those programs that were at once technically possible, politically feasible, and challenging enough to enlist the support of key technical personnel. So that the agency might keep abreast of technical developments, NASA officials thought it necessary to develop capabilities in basic research or in propulsion that were independent of any specific mission or use. This policy lessened the danger, noted in a 1966 Senate report, that "there may be a penalty attached to the 'approved mission' policy for advanced development. Premature obsolescence is one hazard. Commitment of resources before the full cost-benefit is another. The narrowing of component and subsystem engineering is a third." [4]

But the conditions I have listed do not explain NASA's success in managing large-scale technology. Precise goals and organizational flexibility help to set the rules of the game; they define, as it were, a policy space in which NASA could manage its programs. To show how NASA managers worked within that policy space, I want to discuss three areas: the problems faced and met in setting up a headquarters organization; selecting contractors who could operate in the peculiar environment of very large research and development (R&D) programs; and the means by which NASA kept the military at arm's length, while receiving the support necessary to launch and track Mercury, Gemini, and Apollo. These areas, it seems to me, can tell a great deal about the success of NASA's approach to getting its R&D work done. In the final section of this paper, I will mention some of the lessons learned and the extent to which NASA can serve as a precedent for other large-scale endeavors.

Headquarters-Center Relations

Established by Congress in the aftermath of *Sputniks 1* and *2*, NASA quickly grew by accretion, the incorporation of older installations, and the creation of new capabilities into an agency with 36,000 civil service employees and a budget of $5.5 billion by 1965-1966. Indeed, by 1962, NASA had taken on most of the features it possesses today. It was headed by an Administrator supported by a Deputy and an Associate Administrator; together, these officials comprised the agency's top management. Under them were bureaus with agency-wide functional responsibilities for procurement, budget preparation, personnel, public affairs, and legislative affairs. Additionally, there were four program offices, each headed by an Associate Administrator and responsible for NASA's substantive programs. From 1963, these offices were: Space Science and Applications; Manned Space Flight, which was responsible for Mercury, Gemini, Apollo, and the follow-on to Apollo that became Skylab; Advanced Research and Technology, which managed NASA's aeronautical research, as well as the supporting research for the other program offices; and the Office of Tracking and Data Acquisition. All of the field centers reported directly to the program offices. Thus the Marshall Space Flight Center in Huntsville, Alabama, the Kennedy Space Center at Cape Canaveral, and the Manned Spacecraft Center at Houston all reported to the Associate Administrator for Manned Space Flight. The older research centers which predated NASA reported to the Office for Advanced Research and Technology, while the Goddard Space Flight Center in the Maryland suburbs of Washington reported to the Office of Space Science and Applications. There was one other installation that was unique. This was the Jet Propulsion Laboratory in Pasadena, California, which was operated by the California Institute of Technology under contract to NASA. JPL was (and still is) responsible for managing NASA's deep space and interplanetary probes and, consequently, reported to the Office of Space Science and Applications.

Clearly, a summary of names and reporting responsibilities tells very little about relations between headquarters and the field centers. The tension between headquarters and the centers was built into NASA. Headquarters, itself almost a kind of rival installation, had certain key functions: to prepare and defend the agency budget, to allocate funds for

R&D and the construction of facilities, and to serve as a central control point. Beyond this, there were problems which senior management could hope to resolve only after years of trial and error. One of these was whether the centers should report directly to the agency's general manager—the Associate Administrator—or to the heads of the program offices. The first approach was the logical solution when the centers were involved in a variety of projects; the second, when each center had a carefully defined task distinct from the other centers. Another problem was how centers reporting to one office could work with those reporting to another. A third was the problem of project assignment: whether to give the entire project to one center, split it between the centers and designate one as "lead," or put the entire project management team in head-quarters. A fourth problem was how to convert the older research-oriented institutions into managers of large development contracts. And all of these problems were compounded by the difficulties faced by head-quarters and the centers in communicating with each other. The greater the pressures of time, the faster the rate of significant change in the en-vironment; the more interrelated the various programs, the more difficult and necessary adequate communications would be.

Yet, by the end of 1963, all of these problems had been provisionally solved. NASA's top officials stressed that project management was the field installations' responsibility and that, within certain limitations im-posed by Congress, directors and project managers could move some funds from one budget category to another. For all flight projects except Apollo, there was to be one lead center, regardless of how many installa-tions actually participated. The tools for getting the job done would be grouped in related fashion. Thus the Office of Applications, which used the same launch vehicles and centers as Space Sciences, merged with it in 1963. Each center was to have the capacity to manage large development contracts, and, if necessary, assign projects for which new skills would have to be recruited; the skills to integrate the subsystems of a project parcelled out among two or three different centers; and the ability to draw on the resources of other centers instead of duplicating them needlessly. Concurrent with the change by which the centers reported directly to the program offices, NASA instituted two other reforms which greatly im-proved operations. It unified all launch operations at Cape Canaveral, where previously each center had had its own launch team; and it established intensive monthly status reviews, at which Associate Ad-ministrator Robert Seamans would sit down with the heads of the pro-gram offices to review planned versus actual allocations, at the centers and at contractor plants; planned versus actual expenditures; milestones in program and procurement schedules; and advanced studies prior to their completion. These recurring meetings enabled top officials to use overlapping sources of information, give all points of view an airing, and eliminate the middleman in channeling information upward.

NASA Procurement Strategies

Next to the ordering of headquarters-center relations and inseparable from it, the most important decision made by NASA officials was to rely on private industry rather than in-house staff to implement its R&D programs. Contractors were involved at every stage of R&D and for every purpose, from the preparation of advanced studies to systems engineering, manufacture of hardware, checkout of flight equipment, operation of tracking stations, etc. From the outset NASA chose to follow the Air Force and the Atomic Energy Commission in contracting out; in particular, the Air Force and its intercontinental ballistic missile (ICBM) programs were the only programs since the Manhattan Project comparable to the NASA mission. Both ICBM and Apollo had in common technological complexity, tight time schedules, unusual reliability requirements, a general absence of quantity, and little follow-on production. Although some 20,000 firms were working on Apollo in the mid-1960s, a 1969 study showed that NASA had bought only 20 Mercury, 13 Gemini, and 38 Apollo spacecraft including test models and spacecraft modified for changed mission objectives. NASA usually had to contract for products whose main features could not be precisely defined in advance, so that there was no clear-cut basis on which the bidder could make realistic cost estimates. For R&D programs of this sort, NASA waived formal advertising in favor of negotiations with selected bidders.

Viewed in this light, the rationale for an in-house technical staff was to enable NASA to retain those functions that, it has been said, no government agency has the right to contract out, functions enumerated by a former Director of the Bureau of the Budget as "the decisions on what work is to be done, what objectives are to be set for the work, what time period and what costs are to be associated with the work, what the results expected are to be . . . the evaluation and the responsibilities for knowing whether the work has gone as it was supposed to go, and if it has not, what went wrong, and how it can be corrected on subsequent occasions." [5] This, in fact, was NASA's position: that the rapid buildup of the Gemini and Apollo programs precluded reliance on government employees alone; that it was agency policy not to develop in-house capabilities already available in the private sector; that NASA employees were needed for technical direction rather than for hardware fabrication or routine chores; that NASA had developed safeguards for policing its contractors; that it was better to let the up-and-down swings in manpower take place in the contractor, rather than the civil service, work force; and finally, that the practice of using support-service contractors had been fully disclosed to Congress and the Bureau of the Budget. NASA was prepared to go even further. When Congress and the White House began to cut NASA's budget from 1967 on, NASA laid off its own employees at several centers before dismissing contract workers. More remarkable still, NASA's position has been sustained in the federal courts and would seem to have government-wide application.

In the short run, NASA's use of negotiated competition for large R&D contracts must be judged a success. It enabled NASA to assemble manpower—some 420,000 contract and government employees in 1966—and disperse it gradually as the manned space program phased down. It tapped capabilities already available, and saved NASA from having to develop those same capabilities from scratch. Since the largest prime contracts—those for the Apollo spacecraft or the Saturn rocket, for example—required thousands of subcontractors, NASA's R&D monies were spread over much of the United States, so enlarging the agency's clientele. But the system had serious weaknesses. Despite the introduction of incentive provisions and the negotiation of contracts for successive phases of the R&D process—phased project planning—NASA was unable, despite the most strenuous efforts, to police its contractors. The idea behind incentives was to reward the contractor for staying within cost and on schedule and to penalize it for falling short. But while incentives might reduce they could not eliminate the technical uncertainties dogging most R&D programs. A contract designed to cover everything from the early development phases to small-quantity production was not flexible enough for the kind of program where the end item changed over the life of the program. The contradiction between fixed targets and changing programs was not easy to reconcile. Moreover, the sheer size of these programs made it exceedingly difficult to find out what was going on in the field. NASA did not even pretend to review work below the first tier of subcontractors. NASA's inability or unwillingness to force its contractors to make major design changes led to the January 1967 fire which killed three astronauts and caused the Apollo program to slip 18 months.

Another flaw in NASA's procurement system was that competition for major contracts dwindled in the 1960s. There is reason to believe that NASA chose competitively more frequently in the late 1950s and early 1960s than it did later. It may be that by 1965 there were fewer new systems on which to bid, or that the high cost of entry locked out prospective competitors. It was not only expensive to get into the space business but even more expensive to stay in; thus Grumman, NASA's number two prime contractor during the later 1960s, virtually withdrew from space systems after completing its work on the lunar module and the Orbiting Astronomical Observatories, both of which were plagued with overruns and technical difficulties. And as aerospace firms merged or were bought up by competitors, NASA found itself locked into an industry structure for which it was partly responsible.

Finally, even in the 1960s NASA did not have all the in-house skills it would have needed to provide its contractors with complete technical direction. NASA had to call in Boeing to integrate the Apollo spacecraft with the Saturn V launch vehicle; General Electric, to check out flight equipment at Cape Canaveral; AT&T, to set up a wholly-owned subsidiary to do systems engineering and long-range planning for NASA. It must be stressed that NASA, in the Apollo program, possessed a far greater depth of experience and talent than the Air Force's laboratories or the Special Projects Office that developed the Navy's Polaris. NASA personnel determined the conditions under which contracting would be necessary, anticipated problems before the contractor, reviewed the contractor's work, and terminated the contract. But there were areas where NASA engineers did not have the same degree of competence as their contractors, and where NASA had little choice but to accept the contractor's analysis. This was the case when NASA had more than 35,000 employees. In the era of the Space Shuttle, NASA, with perhaps 40 percent fewer employees, probably has less real control, less ability to change the scope of work, than it had 15 years ago.

NASA-Defense Relations

The final area I would like to discuss is NASA's relations with the Department of Defense. Units such as the Defense Supply Agency, which administered many NASA contracts, the Army Corps of Engineers, which managed NASA's largest construction projects, and the Air Force, which detailed officers to serve as program managers and directors of center operating divisions—all of these provided essential support to the agency. This was in addition to the early, once-only transfers of launch vehicles like Saturn, spacecraft like Tiros, contractor-operated facilities like the Jet Propulsion Laboratory, and the technical skills of Wernher von Braun's team of engineers. Simply to list examples, however, gives only the barest hint of the significance, for NASA, of the totality of such support. The essence of the NASA-DoD relationship had far more to do with mutual need than with philosophical arguments concerning the existence or the desirability of one space program or two. The Space Act itself could only outline the scope of interagency relations in the most general way. The act declared that, while aeronautical and space programs would be managed by a civilian agency, "activities peculiar to or primarily associated with the development of weapons systems . . . or the defense of the United States" would remain DoD's responsibility; and it enjoined NASA to make available "to agencies directly concerned with national defense . . . discoveries that have military value or significance." It is well, then, to set aside preconceptions. "Civilian" and "military" were not the same as "peaceful" and "non-peaceful"; duplication of programs could be "warranted" or "unwarranted"; while much of the struggle over the military uses of space was as much between elements within DoD as between DoD and NASA.

The principles underlying the U.S. space program resulted less from anything enunciated in the Space Act than from President Kennedy's decision in May 1961 to assign the lunar-landing program to NASA. But this decision was preceded by earlier moves by NASA and DoD officials and by Congress to prevent an Air Force takeover. Three of these moves were particularly important: the agreements ratified by Webb and civilian Defense and Air Force officials which laid the ground for further cooperation; the March 1961 order of Secretary of Defense Robert McNamara which, by assigning most DoD space programs to the Air Force, thereby gave the Secretary tighter control over all military space operations; and the pressure exerted by the House Committee on Science and Astronautics, which authorized NASA's budget, to give NASA the lion's share of manned space programs. With the backing of the President and much of Congress and the acquiescence of McNamara, NASA, on the one hand, staked out its position as an independent agency while, on the other, waging a quiet behind-the-scenes battle with DoD to maintain that independence. Beginning as an agency heavily dependent on DoD support, NASA succeeded in freeing itself from overt DoD control by 1963. Whether it was the management of Gemini, the management of what became the Kennedy Space Center, or the existence of colocated NASA and DoD tracking stations, the pattern was the same. NASA would cooperate with DoD, but never to the point of giving away its authority to meet its needs. NASA asserted its right to modify military launch vehicles to serve as boosters, let contracts to firms already heavily involved in defense work, and conducted advanced studies on manned space stations at the same time that DoD was trying to develop its own Manned Orbiting Laboratory.

During NASA's first three years, the Air Force went to considerable lengths to become the dominant partner in the national space program. Even some years later the director of NASA's Office of Defense Affairs could observe that "the Air Force is inclined to look upon NASA as a competitor rather than a partner in the field of space." By 1963, however, the Air Force needed NASA almost as much as NASA needed the Air Force. NASA was doing important research in the life sciences and propulsion, and its centers had test facilities that the services needed badly. The framework within which the two agencies had to coexist had to accommodate many arrangements: whether it was a program managed by one agency with the other sharing in the planning of experiments; a joint program; a program started by one agency and transferred to the other; a joint program mostly funded by one agency; or programs whose success depended on the functioning of separate, cooperating systems. The preconditions for cooperation were that DoD accept NASA's definition of a coordinated program as one where concurrence was "not required as a pre-condition to further action" and that both agencies should centralize the organization of their space and launch vehicle programs to make cooperation possible. Between 1960 and 1963, these conditions were met.

The Lessons Learned

The conclusion I wish to draw from these cases is that NASA's remarkable success in managing R&D depended on the ability of the agency's top officials to enunciate goals, to shape the agency from within, to delegate to the program offices and centers the authority to get the job done, and to keep DoD at arm's length. Once NASA began to lose the support of the White House and Congress—roughly from 1967—the difficulty of running the agency became much greater and NASA began to resemble any other large government organization which redoubles its effort as it forgets its aim. The same combination of organizational and political elements which made for success in the first half of the 1960s could not stay the reduction and cancellations extending from 1967 almost to the present.

Timing Matters

In 1961, NASA was still a loosely-structured agency whose field centers worked in relative isolation from each other and from headquarters. The lunar-landing mission demanded much greater coordination—and for the time being, greater centralization—than had been the case. One of the most important aspects of the Apollo program was the speed with which the crucial administrative and program decisions were made and the major prime contracts awarded. Except for the decision to go to all-up testing (the testing of all the major Apollo components together), the principal Apollo program decisions were made between August 1961 and the end of July 1962. Had they been stretched out over a longer period, it seems unlikely that they would have received the support that they did. A comparison between the establishment of the Manned Spacecraft Center (MSC) and the Electronics Research Center (ERC) in Cambridge, Massachusetts, will bring this out. NASA announced the selection of Houston as the site of the former after a brief survey. Yet the creation of the center generated powerful political support; the site itself was well located in relation to Huntsville and the Cape; and the reasons given for establishing a new center were justified in relation to the Apollo mission. In contrast, almost two years elapsed between the decision to establish the ERC and its formal establishment. There was no such consensus as existed in the case of MSC; NASA could not convince Congress or the public that a capability in electronics research was as vital to the agency as one to develop the Apollo spacecraft. The point is that the agency's top officials made the important decisions while there was time to do so. The 1961 reorganization had to be reversed two years later, but it gave NASA management the opportunity to bring the centers under tighter control than before.

The Importance of Flexibility

Another element in the success of the NASA organization was flexibility: flexibility for the Administrator to appoint to excepted positions, to award major R&D contracts without competitive bidding, to reprogram funds within appropriation accounts and to transfer between them, to devise and administer a custom-tailored entrance examination, etc. Examples such as these represent flexibility within the system, not a departure from it; departures from the norm were allowed by Congress, the Bureau of the Budget, and the Civil Service Commission. This flexibility allowed for that ''free play of the joints'' without which institutional *rigor mortis* sets in. The use of excepted positions, for example, served not only to promote employees from within, but also to bring in new blood and to expose NASA to outside influences. Similarly, without the authority to negotiate major contracts, it is unlikely that the lunar landing would have occurred on schedule. Indeed, this authority was probably more important than the introduction of incentive provisions from 1962 on. Incentives were difficult to administer: they required a great deal of manpower and paperwork, the criteria for incentive payments were hard to pin down, and there was a contradiction inherent in fixing targets for changing programs. NASA management might well have awarded development contracts without adding incentive provisions. But it is hard to imagine Gemini, Apollo, or the orbiting observatories becoming operational had the agency been bound by competitive bidding or other rules that would have constrained its ability to choose its sources. The flexibility available to NASA depended on congressional willingness to tolerate practices that the legislature might have disallowed elsewhere. And when that toleration ceased, NASA fell victim to red tape and the bureaucratic tendency to review everything at least twice. By 1969, for instance, it took an average of 420 days to process a contract involving a procurement plan, 3 months for headquarters to review the plan, and 47 days for headquarters to approve a negotiated contract.

Politics and Effective Strategy

NASA management saw its responsibilities in political terms. The agency's top officials took it upon themselves to justify NASA where it mattered most—to the Bureau of the Budget, whose fiscal authorities set the terms of the annual budget request, and to Congress, which had to authorize the entire space program annually. What Harvey Sapolsky has said about Polaris surely applies here: "Competitors had to be eliminated; reviewing agencies had to be outmaneuvered; congressmen. . . , newspapermen and academicians had to be co-opted. Politics is a systemic requirement. What distinguishes programs in government is not that some play politics and others do not, but, rather, that some are better at it than others." [6] Thus the history of NASA from its establishment to the mid-1960s can be charted in terms of NASA's ability to design its own programs, procure its hardware, and support its spacecraft without overt interference from the military. The transfer of the Jet Propulsion Laboratory and the von Braun team to NASA, the 1961 cooperative agreements on the development of launch vehicles, President Kennedy's decision to assign the lunar mission to a civilian agency, and the 1963 agreement by which DoD acknowledged NASA as lead agency in Gemini, all represent stages by which NASA asserted its determination to run the agency as its officials saw fit. Not that interagency relations can be easily categorized. While most relations can be seen to fall into the categories of support, coordination, and rivalry, there were some that did not fit neatly into any category. There were others, like Gemini, that tended to become more like joint programs over time; while a program like the Manned Orbiting Laboratory was, in some ways, competitive with

Apollo, although the former relied heavily on NASA technology and ground support. Nevertheless, without a strong assertion of independence, NASA would have become what the services anticipated on the eve of the Space Act—a research agency supporting military projects.

The political strategies of NASA management were fourfold: to maintain NASA's independent status as an agency doing R&D; to curb outside interference by advisory and coordinating groups; to seek the approval of Congress in actions that the agency was about to take; and to limit NASA's support for other agencies, the better to concentrate its resources on Gemini and Apollo. NASA's relations with DoD are an example of the first type of strategy; its conflicts with the Space Science Board of the National Academy of Sciences is an example of the second; while NASA's position on the supersonic transport—to maintain an essentially supporting role to the Federal Aviation Administration—reflected the desire of Webb and Deputy Administrator Hugh Dryden not to strain NASA resources to the limit. Additionally, Webb dismantled the office that prepared the NASA long-range plan, precisely to avoid premature commitment to something beyond Apollo.

The Centers and Apollo

As mentioned before, NASA was remarkably decentralized for so large an agency. Perhaps it would be more accurate to say that programs such as Apollo or the orbiting observatories could not have been managed without the delegation of authority to the centers and the Jet Propulsion Laboratory—authority to negotiate contracts up to a specified amount, to transfer funds between programs, to start new research tasks without seeking specific authorization, to shift manpower from one division to another. The strategy of senior management was to give the centers what they needed to get the job done, but not so much that their work would lose its relevance to the agency's mission. During the 1960s, the "research" and "development" centers tended to become more like each other; centers reporting to one program office began to work for others; while those centers with a mixture of projects weathered the budget cuts at the end of the decade better than those with one or two large development programs that were phasing down. One of the most important by-products of Apollo was the pressure it placed on the older centers to get into development work. It was not so much a matter of pressure from headquarters as pressure from within the centers themselves that brought about this change. One wonders if the older centers had much choice; had they remained research centers and nothing else, they would very likely have dwindled into insignificance. The centers had, so to speak, to latch on to the coattails of Apollo.

By 1969, most of the centers, particularly Marshall, were in the early phases of a "withdrawal process" brought on by cuts in manpower and funds. The problem of new roles and missions could be alleviated by the centers, but only in part. NASA officials conceded in principle that a less-than-best laboratory might be closed: if it had served its initial purpose; if there was no likelihood that a new role for the laboratory could be found; if the closing down of the laboratory would not leave a significant gap in the national capability to do R&D work. But most of the centers were adaptable and nearly all had gone through at least one reorganization in the late 1950s or early 1960s, moving from aeronautics to launch-vehicle development, or from development work on guided missiles to lunar and planetary probes, as with the Jet Propulsion Laboratory. By 1969, another cycle or reorganization was under way, as facilities that were no longer needed closed down, others were modified to accommodate new programs, while new facilities like the Lunar Receiving Laboratory at Houston became accomplished facts. Yet the more subtle changes in a center's mission could only occur very gradually. And here, it seems, the failure of headquarters to draft a coherent long-range plan left the centers at a serious disadvantage. The advanced studies and task force reports of 1964-1969 were no substitute for a NASA-wide plan. There were too many planning groups, with little coordination between them; a lack of interest among the centers; and the artificial forcing of the planning process by the creation of President Nixon's Space Task Group early in 1969. Still, top management might have done more to bring the process to some visible result inside the agency. In particular, not enough was done to relate substantive programs to any institutional framework.

In sum, NASA thrived during the early 1960s because of four elements within, or conferred upon, the organization: administrative flexibility; the ability of senior management to play the political game on the Hill, at the White House, and before the public at large; the delegation of program management to the field; and the timeliness with which the important decisions were made. But the same elements were not enough to enable NASA to weather the severest test to which any large mission-oriented agency can be put: namely, how to react to the completion of the original mission. It remains to be seen whether the Space Shuttle will be a truly radical departure for the U.S. space program or an example of an R&D program pushed through development long after evidence accumulated that the mission was not an attractive one.

References

1. James E. Webb, *Space Age Management:The Large-Scale Approach,* McKinsey Foundation Lecture Series, Graduate School of Business, Columbia University (New York: McGraw-Hill, 1969), especially pp. 60–65.
2. Webb, *Space Age Management,* p. 61.
3. Charles E. Lindblom and David K. Cohen, *Usable Knowledge: Social Science and Social Problem Solving* (New Haven: Yale University Press, 1979), p. 86.
4. U.S. Senate, Committee on Aeronautical and Space Sciences, *Policy Planning for Aeronautical Research and Development,* 89th Congress, 2nd Session (May 19, 1966), p. 8.
5. Testimony of David E. Bell, Director, Bureau of the Budget, U.S. House of Representatives, Committee on Government Operations, Subcommittee on Military Operations, *Systems Development and Management,* 87th Congress, 2nd Session (June 21, 1962), p. 44.
6. Harvey Sapolsky, *The Polaris System Development* (Cambridge: Harvard University Press, 1972), p. 244.

COMMENTARY

I.B. Holley, Jr.

As a rural New Englander brought up on the prudential ethic, "eat it up, wear it out, make it do, do without," I used to be shocked when I read about the profligate banking practices of the Jacksonian era. I was inclined to look down my nose at an administration that permitted the irresponsible issue of ill-secured bank notes. Then some years ago I read an essay by Joseph A. Schumpeter which put the problem in a whole new perspective. Inflationary emissions of paper in that capital-starved era were not simply a matter of policy, Schumpeter pointed out; they were a necessity. In the 1830s, government, at all echelons, lacked the necessary tools, the bureaucratic apparatus, to impose and enforce regulatory controls—even if it had been decided, as a matter of policy, that such controls were necessary.

As it says in the cigarette advertisements, "We've come a long way, baby." For those of you in the audience who are under 30, it may not be so evident how far we've come in the way of perfecting governmental apparatus just since the beginning of the space age. And I date this from the launching of V-2 rockets by the Nazis in World War II. In a sense, the first 20 years in space is a tale of advancing bureaucratic competence, and each of the papers presented here offers testimony on that theme.

In my commentary on the interesting papers we have just heard, I shall take them in reverse order, beginning with Arnold Levine's.

Mr. Levine draws our attention to NASA Administrator Jim Webb's comment that perhaps the most important byproduct of the whole space endeavor can be found in great leaps forward in the management skills and administrative procedures devised to organize and operate such "large-scale endeavors" as those required to put men on the Moon.

Let me tell you a story to illustrate just how far we've come in perfecting the apparatus of government and business management. Which is to say, how far we've come in our ability to cope with complex scientific and technological problems. As you all know, during World War II one of the major weapons of our Air Force was the B-17—the Boeing four-engine heavy bomber, the Flying Fortress. Obviously it was of the utmost importance to increase the production of these bombers. Boeing brought in other manufacturers and eventually more than 12,000 B-17s were produced. It was an epic achievement.

But turning out bombers is not just a matter of simple repetition, stamping out more and more copies of the same thing. To keep ahead of the enemy, it was necessary to introduce a continuous stream of design changes or modifications. When we tried to introduce design changes on the assembly line, it slowed up and even stopped production. This would never do. So we set up modification centers, some here in the United States, some in the combat theaters. There, teams of workmen patched on modifications as best they could, an additional gun here, an improved escape hatch there. All of these "quick fix" solutions gave us aircraft that were better able to survive in combat, but they also gave us a chaotic mess of nonstandard airplanes. The world was soon populated with maverick aircraft, scarcely two alike. The spare parts problem became a nightmare.

Gradually, however, administrative procedures were devised so that the whole disorderly, nonstandard mess was brought under control. Modifications were injected directly on the assembly line by an orderly system of block numbers so that similar aircraft could be assigned to the same units, effectively simplifying the spare parts problem. Toward the end of the war, the BDV Committee (for Boeing, Douglas, and Lockheed's Vega, the three firms turning out Flying Fortresses) was functioning so smoothly that components fabricated in one plant could be accurately and readily mated to units on the assembly line in another plant.

The public may glow with pride at the thousands upon thousands of combat aircraft turned out, but how many of us give more than passing thought to the impressive managerial and bureaucratic advances which have made possible each new stride forward on the technological front. Arnold Levine does well to highlight this aspect of the NASA story, for the impressive improvements in the art of guiding and controlling "large-scale endeavors" are light-years ahead of our performance during World War II.

I'm only sorry Mr. Levine did not have time to get down into more detail in his paper to illustrate some of the administrative triumphs of which I speak. Let me mention one or two examples.

One of the most impressive aspects of NASA management is the way in which the leaders of the organization managed to elicit enthusiastic cooperation from competing industrial firms. Despite strong proprietary interests and a necessary profit-making orientation on the part of the major contractors, NASA induced them to exchange technical information almost as freely as if they were scholarly members of a scientific society. If you haven't worked in the rough-and-tumble, cut-throat, competitive atmosphere of the industrial world, you may not appreciate fully the magnitude of this achievement. For those of you just entering the space field, let me assure you there are exciting vistas here for further investigation.

Now let me touch briefly upon yet another managerial innovation. In the unforgiving realm of space, extreme reliability is essential. (Remember astronaut Pete Conrad's famous quip on his dismay at recalling how his vehicle was produced by the lowest bidder!) Manufacturers must be held rigidly to the utmost standards of quality, right out to the leading edge in the state of the art. At the same time, NASA must exercize a continual pressure to hold down costs. How are we going to reconcile the inevitable tension between these polarities? At one end we are driving the manufacturer on to better and better quality; at the other we are needling him to hold down costs. To resolve this tension, NASA officials have had to devise a contractual instrument which would encourage and reward improvements while at the same time providing economic incentives for cost cutting.

We make heroes of astronauts—and rightly so—but how much public adulation is there for the NASA contracting officers who hammered out the clauses which made it possible for contractors to improve quality, to hold down costs, and still earn enough to remain viable as a business firm? And in case you think the participating manufacturers all waxed rich on government contracts, think again. Convair Division of General Dynamics Corporation spent a million dollars on its initial feasibility study on the Apollo Moon flight project—four times as much as the government ultimately paid the firm for the job. And this was substantially true for the other participating firms. Martin Marietta Aerospace spent three million dollars and kept 300 people on design studies for six months, and then Martin wasn't even in on the final production order!

While it's easy to be excited by the impressive triumphs of NASA—managerial as well as technological—I don't want to give the impression that NASA had nothing but successes. Mr. Levine gives us a number of fleeting references to the headaches. I want to single out just one for comment.

He praises the merits of decentralization, remarking on the absence of a central planning staff, and the like. Then he goes on to suggest that one of the assets in the early 1960s was the absence of an advisory committee, an absence which "rescued the agency from becoming captive to policies which might seem to be relevant."

What is he trying to say? This comment appears to be a slap at the whole concept of advisory committees. Do advisory committees tend to stultify the organizations they advise and saddle them with irrelevant policies? As one who headed such an advisory committee for 10 years, I am perhaps unduly sensitive. But my experience points all in the opposite direction. The advice proffered is much more likely to be ignored or circumvented. After all, advisory committees only advise, they don't direct. The power of decision still rests with the duly constituted agency head.

One suspects that Mr. Levine turned that phrase with one eye on the President's Scientific Advisory Committee (PSAC). But while I might agree with Mr. Levine in taking a somewhat jaundiced view of PSAC's advice on the Apollo project, I seriously doubt that one episode justifies the implied generalization which seems to condemn advisory committees out of hand.

Now I want to turn to Pamela Mack's interesting paper. Short as it is, it gives us an excellent glimpse into the kinds of problems which beset a great scientific and technological agency such as NASA. Here we have several examples of an organization that is performing at a nearly miraculous level out on the cutting edge of space science, yet seems to be stubbing its toes and falling on its face when it is confronted with some rather typical human and political problems.

Pam Mack offers us a classic illustration of this kind of behavior with her account of the conflicting aims of the Weather Bureau and NASA. The Weather Bureau with limited funds, wanted a reliable, fully tested, and reasonably priced Tiros weather satellite. On the other hand, NASA, with its entirely understandable zeal to push back scientific and technological horizons, kept pushing for Nimbus, a far more advanced weather satellite. Not only was Nimbus immensely more expensive, it was untried and offered no assurance that it would be available when needed.

Clearly, for all its technological triumphs, NASA had a lot to learn about the political dimensions of its job.

There is a nice bit of irony in this situation. Way back in the early days of rocket research, even before NASA was established, some of the scientists who later played leading roles in NASA found the shoe on the other foot. Their funds were sharply limited so they favored the relatively small and inexpensive Aerobee sounding rocket. With it they could stretch their funds, getting many launchings and more tests from a limited number of dollars. The military authorities, on the other hand, favored the big, expensive, but far more capacious Viking, a rocket which was designed as a follow-on to the captured V-2 German rockets being fired at White Sands for research purposes.

Mentioning the V-2 German rocket leads me to some comments about space science. The first point I want to make is that we find it too easy to read history as a success story. When we see NASA and its accomplishments today—an immense organization, with a staff of thousands of highly talented specialists, and budgets of billions—it is easy to forget that only a few short years ago we weren't even thinking about space. I remember some years ago General Charles Bolte, a distinguished division commander in World War II, made a great impression on me when he said, ''Don't study the last battle when you won the war: that's too easy. Study the first battle when you were taken by surprise and you had to fall back. . . .'' Applying that military analogy to space science, I'd like to suggest that perhaps the most fruitful point for study is back in that period before we even recognized the need for a space program. I'd like to tell you a story to illustrate my point.

I was out at Wright Field, the old Materiel Command, then called the Air Technical Service Command, towards the end of World War II. Not long after V-E Day the officers of the command assembled to hear a report on German research and development. Among other things the speaker told us about uncovering German plans for establishing stations in space from which to bomb the United States. The idea seemed so far-fetched, so impossible, that a roar of laughter swept through the hall. But our imagination wasn't ranging far enough! The important task is to conceptualize the challenge clearly. This the Germans did. Then we picked up the ball and ran with it. Would we have launched a space program if they hadn't pointed the way? Clearly our debt to them is great. (Speaking of our debt to the Germans, that reminds me of a story which made the rounds in the early days of the space effort. It seems that a Russian spacecraft would repeatedly encounter a U.S. craft in orbit. Each time the Soviet pilot would greet the American in Russian and the latter would reply in English. Finally, one of them blurted out, "Why don't we cut out this nonsense and speak German?")

That brings me to the second observation I want to make about space science. You'll never understand the scientist's motives if you look to the program justifications they present to Congress and the like when seeking budgetary support. Those are good reasons but not the real reasons. What drives the scientists on is sheer zest for the game. It's fun. It's exciting, and it's immensely satisfying.

Let me conclude these remarks by an observation that relates to all three of the papers. As you have heard, the space age requires an endless array of talents: scientists with creative vision; clever engineers who can cope with intractable problems; imaginative contract negotiators who can reconcile quality and cost; innovative managers who can escape the stultifying constraints of civil service, and so on down through a long list of specialized skills. But above all we need generalists, gifted individuals who can rise above their own specialties to become the commanders, the directors, the administrators of "large-scale endeavors." My unanswered question to you—the audience—is this: How are we going to find these gifted generalists? How can we best develop them? What combination of education, training, and experience will most readily produce this kind of talent—with the least social waste?

LITERATURE AND THEMES IN SPACE HISTORY

THE NEXT ASSIGNMENT:
THE STATE OF THE LITERATURE ON SPACE

Richard P. Hallion

It is both an honor and a pleasure to have been invited to address this conference on the history of space activity. My topic concerns the state of literature on space. It is both a survey of what I believe to be the most worthwhile sources for information on the space age to date, as well as a commentary on the areas of interest that have attracted the attention of commentators and historians. Finally, I attempt to posit some notions of what we should do in the field of aerospace historiography over the coming few years. While not vast, respectable literature on the history of space activity is already large enough to warrant our review. For this reason, symposiums such as this can serve a most useful function in enabling us to take stock periodically of what has been done.

To date, the literature on the space program has broken down into works treating major topics, such as theoretical underpinnings and biographies; survey histories; studies in comparative history; the legal and political aspects of spaceflight; the postwar period through the impact of *Sputnik*; comparative and detailed examinations of the American-Soviet space rivalry; the implications of space for defense; the heroic era of American space exploration; social commentaries on the space program; memoirs of space explorers; and, last but not least, the dreams of futurists. The works discussed in this paper constitute what I believe to be the more significant works in these fields; it is a very personal interpretation, and certainly open for comment and suggestions by others.

The exploration of space is a 20th-century happening made possible by the development of large rocket boosters capable of placing various kinds of payloads into space. The development of this technology involved complex interrelationships between technologists, the scientific community, federal and military research organizations, the national defense establishment, and those charged with responsibility for foreign and domestic policy. It is not a uniquely American story, though the openness of the American space program has aided those historians, social scientists, and practitioners of science and technology who have chosen to examine various facets of space utilization and exploration.

The three major pioneers of the modern space age were Konstantin Tsiolkovskii, Hermann Oberth, and Robert H. Goddard. Tsiolkovskii's writings and notes have been published in Russian and translated as the *Collected Works of K.E. Tsiolkovskiy* in three volumes, edited by Anatoliy A. Blagonravov (NASA, 1965). Oberth's *Wege zur Raumschif-*

fahrt and *Die Rakete zu den Planetenräumen* have been translated and published by NASA as well, as *Ways to Spaceflight* (1972) and *Rockets into Planetary Space* (1965). The American Robert Goddard is the subject of an excellent biography by Milton Lehman, *This High Man* (Farrar, Straus, 1963), that concentrates on Goddard's trials and tribulations, as well as his occasionally mystical and secretive nature. Goddard's own reports, notes, and papers have been published in three volumes, *The Papers of Robert H. Goddard* (McGraw-Hill, 1970), edited by Esther C. Goddard (his widow) and G. Edward Pendray.

The history of rocketry itself is a broad topic, and the literature is vast and mixed in quality. A good introduction to the technology is Eugene M. Emme's *The History of Rocket Technology: Essays on Research, Development, and Utility* (Wayne State University Press, 1964), a series of essays by practitioners, economists, and historians on various topics ranging from early satellite proposals to rocket airplanes and the origins of space telemetry. Bruce Mazlish has undertaken an ambitious comparative study of the growth of the railroad and the emergence of the space program in *The Railroad and the Space Program: An Exploration in Historical Analogy* (MIT Press, 1965), with essays by such noted authorities as Alfred Chandler, Robert Fogel, Thomas Parke Hughes, and Leo Marx, in an effort to study the impact of both the railroad and the space program upon American society.

The exploration of space is not, of course, purely a matter of science and technology. There are also important questions concerning the rights of nations and the conduct of international affairs, as the recent crash of a Soviet satellite in Canada, the well-publicized reentry of Skylab, and concern over space broadcasting and remote-sensing satellites all indicate. A useful introduction to joint efforts in exploration and utilization of space is Arnold W. Frutkin's *International Cooperation in Space* (Prentice-Hall, 1965), which examines the various international considerations that can influence the conduct of technology and science. George S. Robinson's *Living in Outer Space* (Public Affairs Press, 1975) furnishes the perspective of a lawyer on the legal aspects of spaceflight.

Generally, the history of spaceflight can be arranged to reflect four major periods: the early years of large rocketry, beginning in the 1930s, but with special emphasis on German efforts and the immediate postwar years; *Sputnik* and its aftermath, with the emergence of a "space race," and the first utilization of space; the "heroic era" of manned spaceflight, to the landing of Apollo 11 on the Moon; and the post-Apollo years. The single best source book on rocket development in Nazi Germany and the subsequent influence of Wernher von Braun's "Peenemunde team" upon American rocketry is Frederick I. Ordway III and Mitchell R. Sharpe, *The Rocket Team* (Thomas Y. Crowell Publishers, 1979), which is based on copious documentary research supported by extensive oral history interviews. An indigenous and highly successful American effort to build an upper atmospheric sounding rocket is gracefully and wittily treated by Milton W. Rosen in *The Viking Rocket Story* (Harper, 1955), written by a Viking project engineer in the halcyon days prior to *Sputnik*. The first American satellite effort, the Vanguard project, is thoroughly examined by Constance McLaughlin Green and Milton Lomask in *Vanguard: A History* (Smithsonian Institution Press, 1971), including the shattering effect that *Sputnik* had upon the program and its subsequent execution. The turbulence of the immediate post-*Sputnik* era is captured by a memoir of President Dwight D. Eisenhower's, "Missile Czar," James R. Killian, Jr., in *Sputnik, Scientists, and Eisenhower: A Memoir of the First Special Assistant to the President for Science and Technology* (MIT Press, 1977), which casts light on Washington's space politics milieu.

During the troubled days of the early space race, a variety of individuals attempted to study the Soviet space program from afar. Much of the contemporary literature is quite fanciful, but subsequent works have succeeded in generally portraying the origins, goals, and conduct of the Soviet space program with accuracy. A popular and well-written account that is the best journalistic work is Nicholas Daniloff's *The Kremlin and the Cosmos* (Knopf, 1972). Charles S. Sheldon of the Library of Congress has written extensively on the Soviet space program, producing the most authoritative and insightful works, especially his *Review of the Soviet Space Program with Comparative United States Data* (McGraw-Hill, 1968), *United States and Soviet Rivalry in Space: Who is Ahead, and How Do the Contenders Compare?* (Library of Congress, 1969), and *United States and Soviet Progress in Space: Summary Data through 1971 and a Forward Look* (Library of Congress, 1972).

Not all observers were restricted to studying from afar. One of the major developments of the space age has been the emergence of reconnaissance satellites using sophisticated electro-optical sensors to furnish strategic intelligence. Philip J. Klass, a technical journalist, has written perceptively and authoritatively of both Soviet and American "spy satellites" in his *Secret Sentries in Space* (Random House, 1971), including the ways in which such craft influence the conduct of foreign relations, and the basic technological questions involved in their design and employment, as well as the general history of intelligence gathering from space. The transfer of this technology to scientific exploration is highlighted by Merton E. Davies and Bruce C. Murray in *The View from Space: Photographic Exploration of the Planets* (Columbia University Press, 1971), a fascinating historical, technological, and scientific study.

The "heroic era" of American manned spaceflight has been admirably treated by a series of NASA-sponsored histories that are remarkably free of the boosterism that so often afflicts official accounts. These studies are project-oriented, tracing the development of a specific program, but they also examine a number of other factors including social, political, and economic matters. They should serve as a model for all government historians. The American manned space program involved the Mercury, Gemini, and Apollo programs, as well as the post-Apollo Skylab and Apollo-Soyuz Test Project (the latter a joint U.S.-USSR mission). The following can all be recommended without reservation, and constitute just a sampling of the studies that the NASA History Office has sponsored: Loyd S. Swenson, Jr., James M. Grimwood, and Charles C. Alexander, *This New Ocean: A History of Project Mercury* (NASA, 1966); Barton C. Hacker and James M. Grimwood, *On the Shoulders of Titans: A History of Project Gemini* (NASA, 1977); R. Cargill Hall, *Lunar Impact: A History of Project Ranger* (NASA, 1977) (Ranger was an unmanned lunar exploration spacecraft); Courtney G. Brooks, James M. Grimwood, and Loyd S. Swenson, Jr., *Chariots for Apollo: A History of Manned Lunar Spacecraft* (NASA, 1979); Edward C. Ezell and Linda N. Ezell, *The Partnership: A History of the Apollo-Soyuz Test Project* (NASA, 1978). John Logsdon's *The Decision to Go to the Moon* constitutes not only an insightful and important reference on the political environment surrounding the decision to undertake Apollo, but a major pioneering study in analyzing the social, political, and economic impacts upon mid–20th–century technology. A good reference and introduction to the Apollo program and its social, political, technological, and scientific significance is Richard Hallion and Tom D. Crouch's *Apollo: Ten Years Since Tranquility Base* (National Air and Space Museum/Smithsonian Institution Press, 1979), a series of essays by authorities in various fields ranging from space art to lunar geology. Henry S. F. Cooper has written an excellent account of the near-loss of *Apollo 13* in *13: The Flight That Failed* (Dial Press, 1973). Planetary geologist Farouk El–Baz has examined the scientific harvest available from space sensing in *Astronaut Observations from the Apollo-Soyuz Mission* (National Air and Space Museum/Smithsonian Institution Press, 1977). One of the most imaginative aspects of the Apollo program was NASA's art project whereby leading artists were invited to record their impressions of the whole space effort. Two noted artists who were administrators of this program, H. Lester Cooke and James Dean, have collected the reflective and often stimulating results of this project in *Eyewitness to Space: Paintings and Drawings Related to the Apollo Mission to the Moon* (Abrams, 1971).

Norman Mailer has written of what Apollo meant to him and the "Aquarius Generation" in his *Of a Fire on the Moon* (Little, Brown, 1969). Tom Wolfe, in his often zany and insightful *The Right Stuff* (Farrar Straus Giroux, 1979), has examined the world of the test pilot and astronaut, and the occasional tensions between the two. The best participant account of manned spaceflight—and one of the finest aviation memoirs written to date—is Michael Collins's humorous, thoughtful, and lively *Carrying the Fire: An Astronaut's Journeys* (Farrar Straus Giroux, 1974), a recollection of the Gemini and Apollo programs, and a host of other things, by the former command module pilot of *Apollo 11*.

The future of spaceflight is open to a wide range of speculation, particularly as the United States contemplates relatively routine Earth-orbital operations with the NASA Space Shuttle transportation system. What will be the nature of space exploration and utilization in the decades ahead? One glimpse is that of physicist Gerard K. O'Neill's *The High Frontier: Human Colonies in Space* (William Morrow, 1977). O'Neill envisions gigantic, high-technology, cost-effective space colonies orbiting the Earth and bringing almost unimaginable benefits to human society, a view sharply debated by technologists and social scientists alike. Nevertheless, it is useful for the historian to be aware of such works, and to recognize that the space practitioner today may well be regarded as a prophet tomorrow.

This represents but a brief sampling of the relevant literature available on the space program. It is, however, indicative of the topics that have interested historians and observers through the years.

One's first reaction to all this must be how *little* research has actually been done in a serious, scholarly vein on the space program. For example, our best sources on the Apollo program have been a series of histories and works generated by the federal government itself. To the historian, ever alert to the pitfalls of "official" history, it is refreshing, then, to note that these are remarkably frank works, and as historians we should doff our hats to their authors and the agencies responsible, particularly the NASA History Office.

A second reaction might be how little has been written even in a popular vein. Unlike aeronautics, which has been exhaustively examined by scholars and buffs alike, the space program has not produced the same number of popular pieces or respectable organizations claiming to document its history. Thus, many of the basic secondary sources that a historian normally consults before embarking on a detailed research investigation are missing. It must be added, however, that some might well see this as a blessing. One problem faced by historians of aeronautics is the very bulk of the secondary material, and the fact that much of it is buff literature of doubtful value that often acts to hinder and sidetrack the historian trying to mine it for a few rare nuggets.

Clearly there is a serious need for good biographical studies of the principal pioneers—men such as von Braun, Walter Hohmann, and Korolyov. There is, for example, only one decent Goddard biography (that of Lehman), and it is, of course, now out of print. We do not yet understand the workings of the rocket community and rocketeers; biographies and autobiographies and memoirs would go a great distance in removing this deficiency. Fortunately, there is some evidence of a change taking place. The historical sessions of the American Institute of Aeronautics and Astronautics, the American Astronautical Society, and, especially, the International Academy of Astronautics have generated over the last 10 years an increasing number of excellent memoir papers and biographical articles on such individuals as Eugene Sanger and Guido von Pirquet. This is producing some useful raw data, together with insight into the comparative development of astronautics in various nations.

One historian whose work merits special attention is Frank Winter of the National Air and Space Museum, Smithsonian Institution. Winter has generated a number of articles over the last two decades documenting the early history of rocketry from antiquity to the end of the 19th century, unearthing many little-known, yet influential pioneers, and broadening our knowledge of more popular ones such as William Congreve and William Hale. Currently he is completing a study of the early rocket societies in the 1920s and 1930s and their subsequent contribution to the growth of astronautics technology. This study, when complete, should go far in increasing our understanding of how the pre–Second World War "rocket community" flourished, in much the same fashion that Tom D. Crouch's work on early American aviation brought new light to bear on that well-travelled, if little understood, period in aeronautical history.

Survey histories are needed on both the Soviet and American space programs. Information on the former, of course, is less easy to come by than that of the latter, though, thanks to the work of V.N. Sokolskii of the Soviet Academy of Sciences, a surprising amount of research has been undertaken and is now available to the West. A good survey of European rocketry needs to be done. Some popular accounts have, of course, been written on these topics. What is needed is the scholar's touch—thorough research, precise and insightful writing, and the ability to concentrate on the forest of aerospace development as opposed to the trees of individual rockets, missiles, and spacecraft.

One of the problems in the history of science and technology has been the demand that the historian and writer be familiar with the science and/or technology of the subject they are discussing. This is especially true in the history of the space program. Mere economic analysis, which has worked passingly well in, for example, the history of air transportation, is insufficient here. What is needed is familiarity with the craft of spaceflight; otherwise, many of the actions of the space administrators and engineers are incomprehensible at worst and confusing and misleading at best. When, for example, historians examine the ballistic versus lifting reentry question that confronted America's space planners in the 1950s and 1960s, they will have to understand at least some of the mechanics of reentry from space and the problems that confront advocates of these respective systems. Yet, without resolution of this question, the whole structure of America's space program in the 1960s would have been vastly different. There are a variety of questions that await the historian who boldly plunges into the mass of official (and usually technical) documentation awaiting our attention: the space program's impact on modern industrial and governmental management techniques; the relationship between the civilian and military space efforts; the role of innovation and invention in space technology; the impact of the space program on our domestic life and in international affairs; the relationship between aerospace technology and technology as a whole; the ethics of rocketry as weaponry; the philosophical implications of our flight from the Earth. These are but a few. As we move firmly towards the third decade of spaceflight, let us note that the history and literature of the space program can be likened to a rocket just after ignition. The clarity of our perceptions may be still obscured by the steamy blast of contemporary events, but the launch is go, and the promise and challenge of our task remain to be fulfilled.

A SPACEFARING PEOPLE:
KEYNOTE ADDRESS

John Noble Wilford

In less than a quarter of a century—one generation—we have become a spacefaring people and our accomplishments rank among the most incredible in the history of human endeavor. We have set foot on another world. We have looked at our own world from afar, seen it whole, from a cosmic perspective. Our voices and images are carried around our world in an instant by relay stations high overhead in space. Our robot craft have scouted all the planets known to the ancients and landed on the red plains of Mars. Soon we will have spaceplanes shuttling people and instruments—and, yes, perhaps weapons of space warfare—into orbit with astonishing regularity.

It may seem so obvious that we are spacefaring people as to be beyond comment, but the import of it has yet to sink in. It may be the one thing for which our time will be remembered centuries from now. And yet so little intellectual effort has gone into understanding how and why spacefaring came about at this time, why it has evolved the way it has, and where it may be leading us as a nation and a civilization. This conference, on the history of space activity, I trust will be—to borrow a phrase—one small step toward an appreciation of this phenomenon of our time.

First, we must understand what was happening in the 1950s, for this dictated the pace and direction of most subsequent space activities. Technology was advancing to the point where spaceflight was no longer a dream but an approaching reality. The rocketry of World War II, pioneered by the Germans, was being fashioned into the first intercontinental missiles for delivering postwar hydrogen warheads. Communications, navigation and control systems, and electronic computers were becoming more sophisticated by the year. Our economy was strong and aggressive. We and our rival superpower, the Soviet Union, were in a competitive, expansionist mood. So it was not startling that in 1955 both the United States and the Soviet Union announced plans to launch small scientific Earth-orbiting satellites as part of the 1957-1958 International Geophysical Year. As everyone knows, the Soviets got their satellite up first—*Sputnik 1*, on October 4, 1957—and the shock in this country and through much of the world was profound.

We had emerged from World War II as the preeminent economic and technological power and were given to condescending remarks about the backward Russians' inability to make even a decent ballpoint pen.

With *Sputnik*, however, we realized we had underrated Soviet technology and feared we had overrated our own. Nikita Khrushchev, more full of himself than ever, boasted that *Sputnik* demonstrated the superiority of communism over capitalism, and in the Cold War atmosphere of the 1950s such a bold challenge had a riveting effect.

These, then, were the circumstances at the beginning of the space age. They give us the first major theme in any study of space history: *A converging of technologies made spaceflight possible in the 1950s, and the geopolitics of the Cold War made a Soviet-American space race all but inevitable.*

"Might-have-beens" make for interesting historical speculation. What if the United States had launched the first satellite? Wernher von Braun had the rocket and could have done it about a year before *Sputnik*, but was under orders from the Eisenhower administration not to—the first American satellite was supposed to be a civilian operation, and von Braun was working for the Army at that time. Presumably, an American first would not have startled the world as much as *Sputnik* did, for American technological leadership was taken for granted. The impact of *Sputnik*, when it followed, would have been much less, another case of the Russians catching up, as with the atomic and hydrogen bombs. And if *Sputnik* had thus seemed less threatening, would the United States have reacted with the kind of space program that it eventually mobilized? Be that as it may, the *Sputnik* challenge—and subsequent other "firsts" by the Russians—set in motion an American response that shaped an aggressive space program for the short haul, but eventually left it virtually directionless and bereft of clear political and public support.

The American response, in outline, was this: The Eisenhower administration, under considerable public pressure, unleashed von Braun, whose team launched *Explorer 1* in January 1958. The main condition of the American program was that it be civilian, at Eisenhower's insistence, and toward this end the National Aeronautics and Space Administration (NASA) was created later in 1958. Since the Russians gave every indication of planning manned flights in space, the new NASA moved immediately and with little debate to initiate an American man-in-space program, Project Mercury. Many in the administration, including Eisenhower in particular, thought our response extravagant—but it was modest compared to what happened as soon as the next administration came to power.

John Kennedy wanted to get the country moving again, as he said, but the economy was sluggish, the invasion of Cuba at the Bay of Pigs had been a fiasco, and the Russians had jumped farther ahead in space with the launching into orbit of Yuri Gagarin in April 1961. Kennedy asked his advisers what we could do to leapfrog the Russians in space, and following their advice he announced his decision to land a man on the Moon before the decade was out. This was the beginning of the Apollo program. And it was a typically American response. It was optimistic and expansive, America challenged by a foreign threat and a "new frontier," going forth to meet the challenge unburdened by serious doubt as to the ultimate success. Which brings me to the second major theme in space history: *The initial driving force for a strong American space program was not scientific, economic, or romantic, but political—the pursuit of national prestige and power by a new means and in a new frontier. This no doubt accelerated the development of spaceflight capabilities and the attainment of high-visibility goals, but it contributed eventually to a serious mid-life crisis for the American space effort.*

These were the initial challenges and responses that are the stuff of mega-history. I will get to a third major theme later, for it pertains to the present and future. But first, some lesser themes emerge out of the early years of the space age, themes that should be explored by political scientists, historians of science, and others interested in how institutions and policies evolve.

From the beginning, though it did not always seem so to the public, we have had a plural space program. One program is open, highly visible, and civilian-controlled—the NASA program of manned flight, scientific and utilitarian (weather, communications, Earth survey) satellites, and the exploration of the solar system. Another program is military and mostly conducted in secrecy, the Pentagon space program of "spy" satellites and orbital vehicles for military communications and navigation. Though NASA used to get a heftier share, the Department of Defense now accounts for at least half of the annual space spending, with every indication that its share will grow even larger.

Two other space programs are gaining. A majority of NASA's launchings in recent years have been for paying customers, the operators of domestic and international communications satellites. Projections are for increasing commercial space traffic, conceived, developed, and operated outside NASA's domain. In addition, the National Oceanic and Atmospheric Administration has been authorized to develop its own space program, which will handle the operational weather and Earth survey satellites as well as some other "applications" satellites. This is consistent with the policy that NASA is restricted to research and development.

Other conflicts have occurred because of a split between the manned and unmanned space programs. Or, as it is often expressed, between big, showy, expensive projects and the more modest efforts relying on instruments alone. President Eisenhower and his science advisers favored the latter, but the post-*Sputnik* momentum gave exuberant life to the former. As Tom Wolfe has pointed out, the astronauts were our modern Cold War equivalents of the medieval knights who stepped forward to engage in single-man combat with the enemy.

A corollary of the manned-unmanned dichotomy is the uneasy co-existence between scientists and engineers in the NASA space program. At the start, the engineers were up front: they had to build the rockets, design the electronics, and develop all the other systems without which there would have been no spaceflight, manned or unmanned. Engineers thus assumed control of the program and generally pushed manned flight because it was the biggest engineering challenge. Scientists chafed at their secondary role and also feared that the expense of manned spaceflight would drain money away from their own unmanned projects and from other nonspace research.

Another theme of conflict running through the early space age involved nationalism versus internationalism. The initial thrust of our program was nationalistic to the core, but several times in the 1960s, as we were exerting every effort to beat the Russians with a Moon landing, Presidents Kennedy and Johnson made overtures (usually through United Nations speeches) to the Russians to engage in some cooperative space ventures. But as long as there was a Cold War spirit, and as long as the Russians felt they were ahead in space and we wanted to get ahead, hope of international cooperation went nowhere. Only after our Apollo victory, and in the new spirit of Soviet-American détente, was it possible to proceed with the largely symbolic *Soyuz-Apollo* flight of 1975. More realistic and productive cooperative ventures are underway now with the growing European space program.

Now, I want to turn to what I believe is a third major theme of the history of space activity, which is: *The first Apollo landing was, in one sense, a triumph that failed, not because the achievement was anything short of magnificent but because of misdirected expectations and a general misperception of its real meaning. The public was encouraged to view it only as the grand climax of the space program, a geopolitical horse race and extraterrestrial entertainment—not as a dramatic means to the greater end of developing a far-ranging spacefaring capability. This led to the space program's post-Apollo slump.*

This calls for a flashback to the 1960s and early 1970s. While the Apollo program was unfolding, there was the continuing Soviet-American rivalry, to be sure, but also the war on poverty, concern for the environment, the tumult of the civil rights movement, and the Vietnam war and the domestic turmoil it caused. We began to doubt old assumptions of the inevitable good of technology, to doubt the inevitability of progress, to doubt ourselves. This was something fundamentally new to American society. The people who in 1961 said, "yessir, let's go to the Moon and beat the Russians" had become a different people by 1969. The old national innocence was lost, the old cockiness was gone.

In this context, it is not surprising that the Apollo Project came in for much criticism, although it retained strong support in Congress. The space race factor remained strong. Opinion polls conducted during the 1960s are revealing. Public approval of the American space program generally jumped after a successful Russian effort; yet the approval rating was almost unaffected by American achievements. Further, when respondents were given a list of certain government activities and asked which should be the first to be cut out of the budget in the event of a financial crisis, the space program usually appeared on top.

We had been conditioned to think of the space program in terms of the Cold War, which was beginning to seem less crucial to what really counted. The media no doubt perpetuated this attitude, for editors generally viewed every story in those days in terms of whether it meant we or the Russians were ahead. But NASA also played the game, because that was the surest route to the Treasury. And there was that deadline, the end of the decade, that perpetuated the horse race aspects. If we made the deadline, that would be it.

We did, as you know, and then support for the space program all but collapsed. There was the feeling: "We won the war, now bring the boys home." NASA came forward with all sorts of plans for landing men on Mars, building permanent space stations in orbit and on the Moon, and developing a versatile spaceplane. But no one wanted a big space program any more. And the other Moon landings were anticlimactic.

We are building the spaceplane, the Shuttle, but nothing else. Even that was underfunded throughout the 1970s, which was a factor in its many delays and technical problems. Still, it offers the promise of what the space program can be—and probably should have been all along. It is not being built simply to match the Russians; it is far superior to anything for which the Russians have shown any capability. It is being built to take advantage of space not only as an arena of geopolitics, which it still is, but also as a place for many other human activities on many fronts: scientific research, exploration, adventure, commercial pursuits, industrialization, perhaps even colonization.

We are now at the point where, thanks to *Apollo*, whatever its provenance, we can contemplate a broad rationale for going into space—to explore and learn and expand the human potential, to provide services and products for human consumption, to defend ourselves.

So, while we consider and perhaps deplore some of the reasons we went into space in the beginning, it is well to remember that geopolitics was the impetus for the rivalry between England and Spain during the age of seafaring exploration. You know what that produced. So may it be for the age of spacefaring exploration.

COMMENTARY

Sylvia Doughty Fries

Richard Hallion and John Noble Wilford together have given us a fine introduction to the scope and substance of the literature that has been inspired by modern man's first journeys into one of the last known frontiers—outer space.

There is, as Hallion assures us, ample material to begin with. We have the papers and biographical studies of some of the pioneers in space-flight—Konstantin Tsiolkovskii, Hermann Oberth, and Robert Goddard. The international and legal ramifications of space exploration have also received preliminary attention.

The National Aeronautics and Space Administration (NASA) itself has been the source and sponsor of some of this basic literature, or I should say some of the basic histories of the space program. It gives those involved in the NASA History Program some satisfaction, I am certain, to be assured that NASA's own histories are notably reliable for their thoroughness and candor. Among the most useful publications of NASA's History Program may be the regularly updated *Guide to Research in NASA History* and the *Bibliography of Space Books and Articles from Non-Aerospace Journals*, which provide avenues through the forest of space-related materials.

To the participants at the outer edge and to the special sensibilities of such contemporary observers as Tom Wolfe and Norman Mailer has been left the task of evoking the personal and poetic dimensions of the long and solitary drift beyond the Earth's atmosphere. Tempted as we may be to regard such works as *The Right Stuff* and Michael Collins's *Carrying the Fire* as pleasant diversions which add color, as it were, to our canvas of space, these private explorations may be the key to what is most missing from our current literature on space.

Veteran observer John Noble Wilford has had ample opportunity to reflect upon the principal themes that have appeared to dominate our public, as well as literary, coming to terms with the manifold opportunities presented by spaceflight. He is clearly troubled—and others share his concern—about the narrowly geopolitical motivation for our initial ventures into space. The maturing of those ventures into a full-fledged space program can be characterized, according to Wilford, by three "themes."

One of these is the pluralism of our space program—a program carried out not by one agency or institution, but shared by NASA, the Department of Defense, the National Oceanic and Atmospheric Administration, and a number of other "user" groups—public and private—that make use of NASA research and development. One could point as well to the pluralism that characterizes the actual conduct of the space program through grants and contracts, and the pluralism of government oversight and planning for our space undertakings.

A second theme is the constant tension between the advocates of manned spaceflight and those of unmanned spaceflight. Intimately related to this tension is that existing between the scientific community and the engineering community. Expanding scientific knowledge and achieving engineering triumphs may not always be compatible goals in a program that must compete for increasingly scarce resources, and for even scarcer public attention.

The third major theme is perhaps less a theme than what Wilford so aptly calls the "triumph that failed." This was, of course, the first *Apollo* Moon landing in the summer of 1969. One need not have been a total cynic to be struck by the theater of the absurd that placed both the agony of Vietnam and Neil Armstrong's lithe lunar steps on front page, center. The boldness of the Moon landing, the technological achievements it represented, could not be disputed. But both in Southeast Asia and at Tranquility Base, the assault of our material resources on foreign terrain was exceeded only by the uncertainty of our purpose. Or, so it seemed to some observers.

Both Wilford and Hallion have expressed some disappointment in the intellectual effort that has gone into comprehending the significance of the fact that we, and not only we, have become a spacefaring people. To illustrate, Dick Hallion has suggested some questions and topics in need of sustained and thoughtful treatment, including the following:

- Space program's impact on modern industrial and government management techniques;
- Impact of the space program on our domestic life and on international affairs;
- Relationship of aerospace technology and technology as a whole; and
- Philosophical implications of our flight from Earth.

Note that Hallion's questions all start with space technology as a given, seeking to understand the space effort's impact and influence on various other kinds of activities. I would like to suggest that we might also learn a few things by examining the space program as something that not only shapes other things, but is itself shaped by influences not necessarily technological in nature. For example:

- The U.S. space program has not been—nor, perhaps, should it be—immune from political considerations. How, then, has it been shaped by the politics of governmentally sponsored and funded spaceflight? What, in fact, are those politics? Who are the important constituencies, and what is their relative power?
- What has been the relationship of NASA to the scientific community? We are aware of tensions, but why do they exist? One could go to the core, perhaps, with a close study of NASA's Committee on the Selection of Experiments for Space Craft. We are off to a good start with Homer Newell's *Beyond the Atmosphere: Early Years of Space Science* (NASA, 1980).
- Thirdly, it would be instructive to have available a thorough analysis, or better yet, several analyses, of the influence of the *institutional* arrangements of our space program(s) on the nature of those programs themselves. For example, the U.S. space program, as we have seen, is fragmented, or to put it more positively, "pluralistic." What effect has this fact had on the development of our space technologies and their applications? Or, NASA has, as a matter of federal policy, been largely confined to the work of research and development, while the business of applications has been left to other agencies, public and private. Why? And has this separation of developer from user hampered or enhanced the evolution of space technologies?

There is a close relationship between the two concerns expressed by Wilford and Hallion this evening, that is, between the relative poverty of our intellectual efforts to understand the significance of space travel for us and our civilization; and the relative uncertainty of our rationales for a space program as a major, national undertaking. However, we cannot contrive effective rationales for space exploration, try though we might. Effective rationales sustain policies and programs precisely because they are not contrived. They reflect the genuine needs and aspirations of real and important constituencies.

The burden of our space program is that it has had only a marginal audience, and marginal constituencies. Of course there are the aerospace industries, and members of Congress from the states in which those industries are located. But the concern for economic survival in those industries and those states, however legitimate in itself, cannot alone sustain a prolonged national commitment to space exploration. There are other constituencies—astrophysicists, certain kinds of engineers, and so on, not to mention the occasional space warrior or visionary. But these constituencies are scattered, and their combined aspirations have not, thus far, coalesced into a coherent vision comparable to high national purpose.

What makes this burden—the thinness of our space program's audience and constituencies—so troublesome is that it has very little to do with space exploration itself. It is due, rather, to a deep strain in our culture, to our love-hate relationship with modern technology. As a culture we are easily sold on the promise of technology as a tool for social or political purposes. At the same time we have a deep-seated, agrarian unease over technology, mirrored in Frankensteinian or Faustian imagery, and reflected in our fear that a single agency—whether public or private—might acquire the ability to dominate the rest of us with its technological powers. This fear is aided and abetted by our long-standing ideological preference for political power that is dispersed, divided, and balanced as the surest guarantee against tyranny.

What this has meant for our space program and policies has been the "pluralism" which characterizes not only our space effort, but *all* federally sponsored science and technology. Pluralism has, no doubt, spared us from the evils it was intended to prevent: capture of the heavens by the military or by a single commercial behemoth such as American Telephone and Telegraph. But it has also meant that there has been no central rallying point, no broadly inspired focus, around which a large, politically unified and important constituency for space could form. The space age has come to maturity in the United States and, no longer a novelty, it has to compete for support with other well-established public interests. It is past time to do some hard thinking.

Why does it matter whether or not American men and women continue to take that long, distant voyage, and what is their ultimate destination? Or if, as some would prefer, we delegate our space travels to friendly (we trust) robots, would we lose in human satisfaction what we might gain in economies and technical proficiency? And if we grow anxious to-day over every mechanical incursion into the forests and ranges of our western slopes, what might be our final thoughts should we attempt to transform the star-studded night into another horizon of mines and factories?

Ultimately we must come to terms with much more than the possibility of space travel in and of itself. If there is a real incom-patibility—and I suspect there is—between the ideological ligaments and common sentiments that bind us together, and the institutional and political requirements of a national space program, then we must come to terms with that incompatibility.

For some of us the *Voyager* spacecrafts' reconnaissance of Saturn was nothing short of awesome. I still have difficulty grasping the fact of the extraordinary intimacy, as the heavens go, with which we were able to examine Saturn's moons, its many rings, and its atmosphere with the aid of those splendid little craft as they sail inquiringly through the boundless skies. NASA's planetary missions constitute a space journey undertaken for a purpose of enduring value. And there are other, similar purposes, like a rendezvous with Halley's Comet, by which the space program could truly elevate our own age, an age of so many self-inflicted wounds, to one of the more memorable in the unforgiving history of mankind. Such would not be a space program as an end in itself, but a venture common to us all, drawing upon the best of our shared intellectual and spiritual, as well as material, resources.

DOMESTIC AND INTERNATIONAL RAMIFICATIONS OF SPACE ACTIVITY

OPPORTUNITIES FOR POLICY HISTORIANS: THE EVOLUTION OF THE U.S. CIVILIAN SPACE PROGRAM

John Logsdon

One of the most attractive features to me of the U.S. space program as a subject for historical study is its relatively finite nature. While the National Aeronautics and Space Administration's (NASA's) probes and telescopes may be looking outward toward the perhaps limitless edges of the universe, the organization itself has had a life span of hardly a quarter of a century and for all of that time has been very self-conscious about the historical character of most of its activities. It is difficult in general for historians to reconstruct how events occurred and, even more, *why* they occurred; I submit that, while still difficult, it is comparatively easier to undertake such reconstructions for the United States space program, at least in its unclassified aspects, than for almost any other human enterprise of similar scope and historical magnitude. And to top it off, working on space history is one way for those of us without high technical competence to get close to what is (to me at least) the great adventure of my lifetime.

My interest, as a trained political scientist interested in what I call "policy history," is in understanding why governments undertake particular courses of action (which is how I define policy) and in analyzing the institutions and processes through which those courses of action are carried out. I spend little time on the equally fascinating history of technological developments *per se*. In what follows, I attempt to trace the evolution of U.S. civilian space policy and of the institutional framework through which that policy has been implemented. Most of this policy history is uncharted territory for the academic historian, although the 1957-1961 period is more adequately described than the two decades since then, and the groundwork for further analysis has been laid by NASA's continuing program of commissioned and in-house histories.

Government involvement with advanced science and technology has perhaps never been as intense as it has been in the space arena; there is much to record and to contemplate in this involvement. Hopefully, the account which follows can provide some clues to areas for fertile historical analysis.

Space Policy Principles: 1957-1962[1]

There were, of course, space activities within the United States prior to the 1958 launch of America's first satellite, *Explorer I*, on January 31st of that year. The military services, particularly the Air Force, had initiated early satellite projects. The United States had agreed to launch a scientific satellite as part of the International Geophysical Year, and the Vanguard project had been authorized by President Eisenhower to meet the commitment. Vanguard was a second-priority project, explicitly forbidden from interfering with the requirements of the nation's crash missile programs, and did not achieve a successful launch until later in 1958. Even though it was carried out by the Office of Naval Research, it was predominantly a civilian program with limited scientific objectives.

During the 1950s, others recognized the potentials of space. They included individuals within the various armed services, particularly the Air Force, because space activity seemed a logical extension of its mission, and the Army, because in the Wernher von Braun rocket team at the Redstone Arsenal in Alabama it possessed one of the leading groups of rocket engineers in the world and needed to find missions to keep that team at work under Army direction. A few individuals within the civilian National Advisory Committee for Aeronautics (NACA) also were beginning to see that the organization's future might well lie in expanding its activities into space, although NACA leadership did not adopt this posture until after the initial Soviet satellite launch.

Indeed, it was the shock of the Soviet *Sputniks* in late 1957 that galvanized the U.S. debate on space policy and programs. That debate extended from the late 1957 period well into the early years of the Kennedy administration. The policy debate was often acrimonious, with a wide variety of perspectives represented and with strongly held institutional and personal positions. The principles which emerged from that debate and which are described below were not solely, indeed not predominately, the result of some "rational" analysis of the appropriate basis for U.S. space policy; like most other public policies in the United States, they represented negotiated compromises among conflicting interests. Hopefully, they also reflected some sense of the national interest in a new area of human activity.

A fundamental principle of U.S. space policy was that *activities in space could be justified not only by scientific payoff, military or intelligence applications, or potential economic or social benefits, but also by political objectives.* That the first three of these motivations were legitimate rationales for U.S. space activity was established early in the space policy debate. President Eisenhower turned to his newly-established President's Science Advisory Committee for counsel on the appropriate U.S. reaction to *Sputnik*, and those scientists included individuals who saw space as an exciting new arena for discovery. They recommended a program which focused on scientific return; the science advisers were also concerned that space science not divert money away from other fields of science, but rather be planned as a separate part of the overall national scientific effort. Since the beginning of the U.S. program, space science has competed, on one hand, with other types of space activities—particularly manned spaceflight—for funds within NASA and, on the other hand, with other areas of science for a share of the government science budget.

The national security community was quick to sense the potential of space as an important arena for military and intelligence activities, not primarily in terms of active military operations but rather in terms of using space technology to perform necessary military support functions, such as communications, navigation, and weather forecasting, and surveillance functions central to strategic intelligence. There was little question from the start that, when space offered a more efficient or a unique way of achieving a military objective, the Department of Defense (DoD) would be authorized to carry out military-oriented space projects. The debate in the early years arose about the limits of legitimate military objectives in space, since the most visionary among the military were suggesting "space planes," manned orbiting stations and lunar missions, strategic interplanetary forces, and other expensive and "far-out" projects as appropriate military undertakings.

The capability to operate in space was also recognized early on as having the potential to lead to applications with both social and economic benefits, and this potential was seen as a legitimate justification for exploratory programs to investigate various applications. In particular, the potentials of space technology for meteorological observation and for relaying communications were recognized as areas of early payoff, and rapidly pursued.

The most vigorous area of debate in the early years of the U.S. space program was over whether strategic political objectives such as national prestige ought to be pursued through space activity. The Eisenhower administration explicitly rejected the idea of using large space technology projects to compete in symbolic, prestige-oriented accomplishments with the Soviet Union; Eisenhower insisted on a policy of "calm conservatism" with respect to the political uses of space technology. This policy was reversed by President Kennedy in May 1961, with his commitment to a man landing on the Moon "before this decade is out." Kennedy was straightforward in his rationale for Apollo; as he said in the speech announcing his decision, "no single space project in this period will be more exciting, or more impressive to mankind." The memorandum prepared by Kennedy's advisers which recommended the lunar landing mission to him was even more explicit, arguing that "our attainments [in space] are a major element in the international competition between the Soviet system and our own. The non-military, non-commercial, non-scientific but 'civilian' projects such as lunar and planetary exploration are, in this sense, part of the battle along the fluid front of the cold war."[2]

A second principle of U.S. space policy, also established by President Kennedy, was that *the United States should be preeminent in all areas of space activity, particularly so in those areas involving the demonstration of technological capability.*[3] In addition to reversing Eisenhower's policy of not undertaking space activities for political objectives, Kennedy also accepted the recommendation that the United States aim for across-the-board supremacy in the development of space capabilities. Apollo was just the capstone of this commitment to preeminence. At the same time as he approved the lunar mission, Kennedy also agreed to a general acceleration of the development of U.S. space technology in booster development, nuclear rocket propulsion, communication satellites, and meteorological satellites. The emphasis in this strategy was on technology development, rather than a program balanced among scientific exploration, socially useful applications, and major technology projects.

A third guiding principle for U.S. space activities was that *civilian and military space activities would be carried out in separate institutional structures.* In the early stages of the debate on space policy, the military tried to build a case for a single national space program under military control; a similar claim reemerged, in muted form, in the early months of the Kennedy administration. However, both Congress and President Eisenhower quickly became convinced that there should be an explicit and clear separation between the civilian space activities of government and those aimed at military objectives. This conviction was reflected in the Eisenhower administration's proposal for organizing the national space program sent to Congress in 1958, and it was never seriously questioned during congressional debate. Nor was President Kennedy receptive to the notion of integrating military and civilian space activities in a single agency, although such a suggestion was made as he assumed the presidency in 1961. As intelligence programs using space technology developed, they were carried out under yet another institutional framework, and as civilian space applications reached the operational stage, they were assigned to a mission agency within the government or transferred to the private sector. Further, NASA, as the civilian space agency, was limited to research and development work related to civilian applications of space technology; the R&D necessary for military and intelligence missions was carried out under the sponsorship of those agencies, rather than using NASA as a single R&D agency for all government space programs. Thus, from the start, the principle of plural space programs rather than a single government program embodied in a single institutional structure was established.

The decision to carry out the government's space activities in a plural institutional context implied the need for some form of effective coordination among separate programs and for some means of developing either mutually consistent space policies for each program or a single integrated national space policy. A primary concern was whether space policy development required a distinct high-level mechanism reflecting its status as a presidential issue, or whether policy coordination could be accomplished through the normal operations of the Executive Office. Various mechanisms for program coordination between defense and civilian space activities were established because of the recognition that, if there were to be no central space agency, some such means were required to insure that there were no unwarranted duplications or overlaps in the various parts of the federal space effort.

A fourth space policy principle was that *NASA would be limited to research and development activities only; NASA would not operate space systems.** The notion that NASA was to be an R&D agency only was incorporated in its organic act, and whenever a question of whether NASA's mandate should be extended to include at least the early operation of a fully developed space applications system has been raised, the decision has been that NASA was required to transfer to some other entity any technology which had reached the operational stage.

A fifth principle of U.S. space policy was that *while the government would actively encourage private-sector uses of space technology, the government would also sponsor research in areas of potential commercial applications in space, both to accelerate the development of those applications and to prevent private monopolies based on space technology.* This policy took several years to evolve. The forcing issue was the desire of American Telephone and Telegraph (AT&T) to invest its own corporate funds in the development of a communications satellite, if only the government would·agree to launch such a privately developed piece of hardware.[4] The government monopolized the capability required to launch payloads into orbit, and that capability had been developed at public expense. For this and other reasons, there was controversy from the

---

* This principle applies particularly to the space applications area. Space science is, almost by definition, exclusively an R&D activity. NASA has, to date, acted as the operational agency for launching nonmilitary payloads into space.

start over the notion of government assistance to a single corporation* in achieving, if not a monopoly, at least a strong initial advantage in the exploitation of space communications.

The Eisenhower administration was willing to leave research and development specifically related to civilian communications satellites to the private sector, but this policy was reversed in the early years of the Kennedy administration. Not only did the government take the initiative in establishing an entirely new entity, the Communications Satellite Corporation (COMSAT), to be the U.S. actor in operating international commercial space communications systems, but the President also authorized NASA to invest public money in communications satellite research and development, thereby helping firms other than AT&T to gain competence in this area without large commitments of their own resources.

A final principle of U.S. space policy was that, although the 1958 Act specified that NASA might "engage in a program of international cooperation," *international cooperation was second in priority to nationalistic objectives and was to be pursued in the context of broader U.S. domestic and foreign policy goals.* Both Presidents Eisenhower and Kennedy saw the potential for space being an arena of substantial international cooperation; this was one rationale offered for placing the U.S. effort primarily under civilian control. However, President Kennedy, by setting preeminence in space technology as a high-priority policy goal, implicitly relegated international cooperation to a lower priority than competitive, nationalistic motivations for the U.S. space program.

These six principles formed the policy framework within which at least the first decade of U.S. space activity took place. They were also the policy principles upon which an elaborate institutional structure for the national space program was developed. The main features of that structure are described below.

Institutional Evolution of the U.S. Space Program

Institutions are created, at least ideally, to embody a particular set of policy choices. As policies change, institutions either adapt, are modified by external forces, or become obsolete. Although the basic institutional structure of the U.S. space program has remained stable over the past two decades, there has been a good deal of organizational adaptation.

---

* Even one, like AT&T, which already had a virtual monopoly on long-distance transmission of voice and video communications.

Whether the changes are adequate to current space policy directions is very much a live question today.

Separate Programs, Separate Structures

The policy decision with the most direct impact on the structure of the U.S. space program was that calling for institutional separation within the government of the civilian and military space activities. In the immediate post-*Sputnik* period, when it was evident that some accelerated response to the Soviet space accomplishments by the United States was required, there were a number of contenders for the job of managing the national effort. They included:

- a single agency for all government space programs managed by the military, either at the level of the Secretary of Defense or by one of the armed services, most likely the Air Force;
- a new cabinet-level department of science and technology which, among its other responsibilities, would have charge of the civilian space effort;
- adding space to the responsibilities of the Atomic Energy Commission;
- expanding the responsibility of the National Advisory Committee for Aeronautics to include a substantial component of space activities;
- creating a new civilian agency with a responsibility for government space activities, except those primarily associated with defense applications (which would be managed by DoD).

It is beyond the scope of this paper to detail the debate which led to the choice of creating a fundamentally new civilian space agency, although one arose around a core of technical capability transferred from NACA.[5] Once the decision to separate civil and military space activities was made, the claims by the Department of Defense and by the armed services that they were the appropriate managers of the national space program found limited political support either within Congress or in the public (outside of those constituencies with close connections to the military). The idea that the U.S. space program in its civilian aspects should be an open, unclassified effort was widely accepted among those concerned with shaping national space policy.

As the government agency concerned with aeronautics research, NACA mounted a campaign to have space added to its activities. However, NACA was an introspective, research-oriented agency with little orientation toward major technological enterprises. Further, it was an agency managed by a committee, not by a single executive; this was an administrative arrangement strongly preferred by the scientific community as a means of insulating from "politics" government activities with strong scientific components. A similar form of organization had been accepted for the Atomic Energy Commission and had been proposed for the National Science Foundation, but was vetoed by President Truman. What President Eisenhower's administrative, budgetary, and policy advisors wanted was an agency responsive to the policy directions of the President, headed by a single individual responsible for implementing those policy directives, and with the capabilities for carrying out potentially major research and development activities. Those activities, it was thought, would be carried out within the aerospace industry under government contract rather than "in-house" with federal laboratories. They thus concluded that the creation of an essentially new federal structure for space, but one built around the NACA core of technical capability and research institutions, was the appropriate route to go.

In the National Aeronautics and Space Act of 1958, the primacy of civilian objectives in space was stated: "It is the policy of the United States that activities in space should be devoted to peaceful purposes for the benefit of all mankind"; and the responsibility for those activities was given to a civilian agency: "Such activities shall be the responsibility of and shall be directed by a civilian agency exercising control over aeronautical and space activities sponsored by the United States. . . ."

One area of controversy in the development of the 1958 Space Act was whether the new space agency should be responsible for all space R&D, including that ultimately to be used by the military for defense applications. The decision was to make explicit from the start the total separation of these two major categories of space activities, with NASA having no direct involvement in military work. Thus the Space Act also declared that the Department of Defense should have responsibility for "activities peculiar to or primarily associated with the development of weapons systems, military operations, or the defense of the United States (including the research and development necessary to make effective provisions for the defense of the United States)."

The formal separation of the civilian and military space activities into different institutional frameworks meant transferring to the new civilian space agency capabilities related to its mission but under military control and, particularly after NASA had been assigned the lunar landing mission, developing new capabilities required to carry out an active space

R&D effort. Within the Department of Defense there was a need to develop a space R&D and a space operations structure, and to determine the division of responsibility between the level of the Secretary of Defense and the various military services. Both the NASA buildup and the development of the initial military structure for space were accomplished by the early 1960s.

Within the first two years of its existence, NASA had transferred to it a number of facilities, programs, and people that had formerly been operating under military auspices. These included, from the Army, the von Braun rocket development team at Huntsville, Alabama, which became the core of the Marshall Space Flight Center, and the Jet Propulsion Laboratory at the California Institute of Technology. NASA was authorized to develop several new field centers related to its mission, including the Goddard Space Flight Center for science and applications programs and the Manned Spacecraft Center (later the Johnson Space Center) for manned programs, and to develop a civilian launch facility at Cape Canaveral, Florida (later the Kennedy Space Center).* These were added to the three former NACA centers: Langley, Lewis, and Ames. In addition, smaller NACA facilities at Wallops Island, Virginia, and Edwards Air Force Base in California came under NASA control. By 1962, NASA had in place an impressive institutional capability, one fully mobilized for meeting a broad set of national objectives in space.

This government institutional base for civilian space programs was reinforced by the development of an elaborate external network of organizations—industries, universities, and nonprofits—involved in carrying out the civilian space program under NASA contracts or grants. (As space activities matured, other government agencies, including the Departments of Agriculture; Commerce; Energy; Health, Education, and Welfare; and Interior also became involved in space-related activities.) At the peak of the Apollo program in fiscal year 1965, fully 94 percent of NASA's budget obligations went to external grants and contracts, and NASA's prime contractors in turn created a wide base of more specialized subcontractors. Of direct NASA procurements in that year, 79 percent went to business firms, 8 percent to educational institutions, 12 percent to other government agencies, and 1 percent to nonprofit organizations. This pattern has remained consistent over the years; in fiscal 1978, the same percentage (94%) of NASA's budget went to extramural procurement, and the distribution among performers was similar—business

---

* There was already a military launch facility at Cape Canaveral.

(81%); educational institutions (12%); nonprofits (1%); and other government agencies (6%).

As the development of government space activities during the 1960s and 1970s continued, the separation between the three components of government activity—civilian, military, and intelligence—became quite pronounced. The government developed and maintained separate and distinct institutional structures for each function, not only in terms of line agencies within the executive branch, but also in terms of policy review, budget development and review, and congressional oversight. There was coordination among the elements of the government space program, but it was limited in scope in comparison to the separate momentum developed by each element of the government space effort.

The NASA structure created by its first two administrators, Keith Glennan and James Webb, has remained basically unchanged during the past two decades. NASA Headquarters in Washington is responsible for policy development, overall management, and technical direction of the various components of the civilian space research program. Technical management of those specific projects is assigned to one of the various NASA field centers. NASA has adopted the "Air Force model" of agency-contractor relationships, in which most R&D work is performed outside the government by the aerospace industry. The government role is that of program and project initiator, technical monitor of contractor performance, and user of the results of the R&D efforts.

The set of field centers under NASA authority today is the same as it was during the early 1960s.* Because NASA is responsible for civilian space activities aimed at a number of different purposes, including science, applications, and development of technological capability, and because the responsibility for each of those missions is lodged in a different field center, one of NASA Headquarters' major responsibilities is allocating priorities and resources across the NASA institutional complex. The vitality of various field centers is closely related to the priority assigned to particular types of space activities under that center's control, and thus there is strong institutional motivation to compete for particular emphases within the overall NASA program.

It may be useful to mention the structure for space policy within Congress. After creating two temporary select committees to deal with space policy in early 1958, later that year Congress established two new

---

* Except for the brief period during which NASA also had an Electronics Research Center in Cambridge, Massachusetts.

standing committees to deal with civilian space matters. In the Senate this responsibility was given to the Committee on Aeronautical and Space Sciences; in the House, to the Committee on Science and Astronautics. Both of these committees derived their visibility and status within Congress from the importance of the programs they oversaw and their authority over those programs. As long as the civilian space program was a matter of high national priority with major budgetary supports, there was a corresponding degree of status in being involved with these two congressional committees. However, as the resources allocated to civilian space activity declined after Apollo, Congress viewed space activities as just one among various science and technology programs of government, and during the 1970s committee jurisdictions and names were modified to reflect this reality. Now NASA and the National Oceanic and Atmospheric Administration (NOAA) programs are reviewed in the Senate by the Subcommittee on Science, Technology, and Space of the Committee on Commerce, Science, and Transportation; there is no separate Senate space committee. In the House, the Committee on Science and Astronautics in 1974 was renamed the Committee on Science and Technology and its jurisdiction was broadened to cover most civilian science and technology activities, rather than being focused primarily on NASA efforts.*

In summary, then, the policy principle of separate civilian, military, and intelligence space programs has resulted in the development of separate and well-established institutional structures aimed at those three objectives. As the priority given to military applications of space has increased, the Department of Defense structure for carrying out those activities has become more elaborate. However, as the priority assigned to civilian space activities has changed, there has not been a corresponding modification of the basic NASA institutional structure or institutional style, although the size of the NASA work force and supporting network of contractors has diminished.

This institutional base offers the potential for rapid mobilization if the nation were to decide to accelerate the pace of its civilian space effort. The consequences of allowing the NASA and contractor institutional bases to shrink further are unclear. It may be a sound national investment to maintain a strong institutional capability within the government for civilian space development, even though that capability is not always being fully utilized. On the other hand, it may also be appropriate, as U.S.

---

* Military and intelligence space programs are authorized by other committees in both House and Senate; this reinforces the separate executive branch structures for the three components of the U.S. government space program.

activities in space mature, to shift more of the responsibility for program and project planning and development to the private sector, with a parallel diminution of government's institutional involvement.

In 1977–1978, under the direction of a National Security Council Policy Review Committee, a major review of the structure of the national space program was carried out. That review validated the fundamental principle of separating civilian and military space activities. It concluded that "our current direction set forth in the Space Act in 1958 is well-founded" and that "the United States will maintain current responsibility and management among the various space programs." [6]

Policy and Program Coordination Required

The decision to separate civilian, military, and intelligence space activities led naturally to the requirement for policy and program coordination among those separate programs. The type of policy coordination needed and mechanisms for coordination have been, and continue to be, controversial issues. The nature of coordination at the program level has been less problematic, and working-level cooperation between civilian and military space efforts has been the rule. However, occasional disputes have arisen over, for example, proposed civilian uses of technology developed for national security purposes.

During the 1958 debate on space policy, a major congressional concern was the relationship between military and civilian objectives in space and some broader set of national interests. Senate Majority Leader Lyndon Johnson, in particular, was convinced that space policy ought to be the subject of presidential attention; the Eisenhower administration was far less convinced that space policy deserved such high priority. Johnson wanted to effect high-level policy coordination by creating an Executive Office mechanism modeled on the National Security Council but dedicated specifically to aeronautical and space activities. The Eisenhower administration reluctantly accepted Johnson's notion as a price of getting the space legislation through Congress, and a National Aeronautics and Space Council was established by the Space Act of 1958. The Space Council was to be a high-level advisory body, chaired by the President and consisting of the heads of other agencies concerned with space activities and several nongovernment members.* It was to assist and advise the President in developing a comprehensive program of aeronautical and space

---

* These nongovernmental members were never appointed and the positions were eliminated when the Space Council was reorganized in 1961.

activities, in assigning specific space missions to various agencies, and in resolving differences among agencies over space policy and programs.

Although the Eisenhower administration agreed to the inclusion of the Space Council in the legislation setting up the national space effort, it never used the mechanism. Rather, space policy under Eisenhower was developed through National Security Council and Bureau of the Budget channels. Eisenhower believed that civilian and military functions in space development were "separate responsibilities requiring no coordinating body." [7] Thus, in 1960, he asked Congress to abolish the Space Council.

This proposal was sidetracked by Lyndon Johnson. When Kennedy won the 1960 election, with Johnson as his Vice President, the new President was convinced to keep the Space Council, but to change the legislation so it would be chaired by the Vice President. During the Kennedy administration, the Space Council hired its first staff members and played an active role in developing the national policies which led to the Apollo program and the administration's position on communication satellites. During the rest of the 1960s, under the Johnson and Nixon administrations, the Space Council continued to exist, but at the margins of most space policy debates. It developed a relatively large (for the Executive Office) staff under the leadership of Vice Presidents Hubert Humphrey and Spiro Agnew. However, as the priority assigned to civilian space programs continued to decrease and as the separate space activities of the government pretty much went their own ways, the Space Council became rather a moribund institution, and in 1973, President Nixon proposed its dissolution. Congress raised no objection and the Space Council went out of existence.

Without a central policy coordinating mechanism during the 1970s, stresses among various government space activities developed. Several of these were the results of disagreements between NASA and DoD over the appropriate national security constraints to be applied to civilian space efforts, particularly in the Earth-observation area. NASA-DoD relationships with respect to the Space Shuttle program have been another area of controversy. It was these stresses, more than any other single influence, that led to the Carter administration review of national space policy begun in 1977.

A major result of that review was the reestablishment of a presidential-level policy review process for space. This process exists in the form of a Policy Review Committee (Space), operating under National Security Council auspices, but chaired by the Director of the Office of Science and Technology Policy. This committee provides a forum for all involved federal agencies (including departments such as Interior and

Agriculture) to air their views on space policy, to advise the president on proposed changes in national space policy, to resolve disputes among agencies, and to provide for rapid referral of space policy issues to the president for decision when required. Unlike the Space Council, the Policy Review Committee (Space) does not have a standing professional staff structure. Rather, it is a recognition of the need to formalize the channels of interaction among the various components of government space activity rather than have policy and program disputes settled through the budgetary review process or other means of interagency coordination.

The structures for coordination among military and civilian space efforts at the program level have had a rather different history than those for policy level coordination. The 1958 Space Act created a mechanism for coordination at this level, the Civilian Military Liaison Committee (CMLC), but that statutory committee, like the Space Council, was a congressionally-imposed structure and was seldom used. Rather NASA and DoD set up a number of working-level groups on issues of interest to both agencies as the early years of the space program passed. The CMLC was eventually abolished and replaced by a non-statutory Aeronautics and Astronautics Coordinating Board (AACB), which formalized the contacts between NASA and DoD at the working level. The AACB was established by a 1960 NASA-DoD agreement and was given responsibility for coordinating NASA and DoD activities so as to "avoid undesirable duplication and . . . achieve efficient utilization of available resources" and undertake "the coordination of activities in areas of common interest." The early years of the AACB were quite productive in terms of data exchanges and creating an awareness of what the other agency's plans were; the AACB continues to exist today as the primary mechanism for addressing major program issues of interest to DoD and NASA in space. However, as the separate NASA and defense programs became more institutionalized in the 1960s and 1970s, there has been a tendency for coordination between the programs to be defensive in character, i.e., aimed at protecting each agency's own programs and "turf."

Putting Research Results into Operation

In the 1958 debate over space activities, the notion of operational civilian space systems did not receive much attention. The Space Act gave NASA the responsibility for most aeronautical and space activities but defined those activities as: (1) "research into . . . problems of flight within and outside the Earth's atmosphere"; (2) "the development, the construction, testing and operation for research purposes of aeronautical and space vehicles''; and (3) "such other activities as may be required for the exploration of space." This language seemed to limit NASA to R&D activities, and that was the general understanding of the agency's mission at the time.

In one area, providing launch services to a variety of customers including other government agencies, COMSAT and other private sector firms, and other countries, NASA has gone beyond R&D to a clearly operational role. Restriction to R&D has had little impact on NASA's efforts in space science and exploration or technology development, but it has had a definite impact in the space applications area.

Limiting NASA to the R&D part of the job of bringing space applications into being means that other users of space technology are necessarily involved in the total application effort. NASA has developed an orientation towards "technology push" efforts rather than a tradition of close coupling with potential users of space technology who would exercise "demand pull" on the development of space applications. While NASA has almost from its start included "technology transfer" functions in its organization, many observers think that NASA has so far done an inadequate job of marketing its technological capabilities to potential users of space application systems.

While an emphasis on developing and demonstrating new technical capabilities is often necessary to convince potential users of their value, especially in situations where no preexisting user community exists, most observers believe that NASA, particularly in its early years, put more stress on pushing the technological frontier in space applications than on developing technology either in response to user demand or in anticipation of the kinds of demands likely to arise as new capabilities became known. In addition, NASA has a history of emphasizing the development of constantly more sophisticated technology in its application programs rather than concentrating on bringing an adequate applications system into early operation. This is at least in some measure a reflection of the institutional reality that, once NASA completes R&D for an applications program, it must transfer that program to some user outside of the agency. There is an organizational tendency to attempt to hold on to programs, even if that means prolonging the R&D phase beyond the socially optimum point.* Since the early 1970s, NASA appears to have put a higher priority on developing closer relationships with potential users of

---

* There may be, of course, technical and managerial as well as institutional reasons why the development of a space application may take longer than originally hoped for. Some also suggest that there have been instances of premature shifts from R&D to operational status in space applications.

space technology, particularly in the remote sensing and advanced satellite communications areas.

The first test of NASA's bias towards continuing R&D in applications was in weather satellites. In the early 1960s, NASA's initial meteorological satellite program, which had been transferred from DoD, was called Tiros. As the agency in charge of space R&D, NASA regarded Tiros as only the first step in weather satellite development and wanted to go immediately to the creation of an advanced meteorological satellite called Nimbus. The Weather Bureau within the Department of Commerce, a potential user agency, had another point of view. Even this initial weather satellite would markedly improve its services, and the Weather Bureau wanted NASA to focus on Tiros rather than initiate a new weather satellite program. However, it took several years and substantial bureaucratic conflict before NASA was willing to shift its emphasis away from the advanced Nimbus development program back to completing Tiros and bringing it to an operational state.[8] Eventually, NASA worked out an effective agreement with the Weather Bureau both to support ongoing meteorological satellite activities and to continue R&D on advanced sensors relevant to meteorological applications.

The complex history of the use of satellites for remote sensing of land and ocean areas demonstrates the institutional problems stemming from, among other sources, NASA's focus on R&D and its lack of close links with potential users of operational space systems. The debate over the appropriate development pace and management structure for the Landsat system has extended over a decade. A presidential decision to assign the operational responsibility for remote-sensing programs to NOAA has provided only a partial resolution of the institutional aspects of that debate.

A major issue as arrangements for operational land remote sensing have been debated over the past decade is whether NASA's charter ought to be revised to extend its authority to the operation of space applications systems. The presidential directive of November 1979 ended this debate with the decision to keep NASA as an R&D agency in remote sensing and to assign civilian Earth observation operations within the government to NOAA, even though there were other claimants, such as the Departments of Interior and Agriculture, to a share of the operational remote-sensing role. Throughout the Landsat program, NASA has emphasized the experimental nature of the early remote-sensing satellites. While it has worked with potential users to make them aware of possible applications of *Landsat* data to their programs, it has also proposed more advanced sensors for orbital evaluation in later *Landsat* satellites. But it has not given priority attention to developing the ground segment, including

associated data management and information processing and dissemina-
tions systems, required for early deployment of a first-generation operational remote-sensing system.

Public Sector-Private Sector Relations

NASA's relationships as an R&D agency for space with other potential users of space applications are relatively underdeveloped; this is particularly the case when those users are not other government agencies, but
rather private sector, profit-oriented firms. The appropriate division of
responsibility between public and private organizations for research and
development oriented towards commercial applications for space
technology has been problematic since the start of the space age.* The
area in which this issue initially surfaced is communications satellite
research. The Eisenhower administration recognized that communication
via satellite was an area of potential major economic payoff and decided,
in keeping with its general pro-business orientation, that
communications-satellite research should be left to those interested in
making a profit in the area. Others, however, feared that allowing only
private entities to develop the technology of space communications meant
in effect giving a virtual monopoly in that area to the corporation with the
most resources available to invest in communications satellite research,
AT&T. From the perspective of those interested in preventing monopoly
power in new areas of human activity, such a development was not
desirable. The situation was further clouded by the recognition that, even
if AT&T or another private entity developed a communications satellite
using its own funds, it would have to depend on a launch capability
developed with public money to place that satellite into orbit. Thus the
Kennedy administration reversed the Eisenhower policy of leaving communications satellite research to the private sector; President Kennedy
authorized NASA to conduct a vigorous program of research in the communications satellite area.

In 1961 and 1962, as an initial space communications capability approached reality, there were those who thought that the government
should not only be involved in communications satellite R&D and make
the results of that research available to a variety of potential private sector
firms for commercialization, but also that the government itself should

---

* Of course, this problem is not limited to the space sector. The issue of federal policies affecting
private-sector innovation, including direct support of civilian R&D, has been a subject of much recent
discussion within both the executive branch and the Congress.

take advantage of that research and undertake the operational satellite communications role, returning the eventual profits to the Treasury. The advocates of this position were not able to gather majority support in the 1962 debate over communications satellite policy. With the creation of a new institution, the Communications Satellite Corporation—which had some aspects of public control, but was fundamentally a new private enterprise—the notion that the government should go into the communications satellite business itself disappeared.[9]

The precedent established during the communications satellite debate was that developing new applications of space technology with commercial potential and nurturing them to operational status is a mixed private sector-public sector responsibility, with the appropriate division of roles to be determined on an ad hoc basis for each area of applications; the goal, however, is eventual private sector operation of space applications systems. In each area in which a space application has reached or approached maturity, such as point-to-point communications and some applications of remote sensing, business structures have emerged which operate as commercial enterprises related to that application. The government has continued to fund research in other areas of space applications with potential commercial utility, including space transportation, materials processing, and other aspects of remote sensing, with the hope of discovering whether there are indeed profitable opportunities for private sector involvement in those areas, and demonstrating to potential operators what those opportunities are. It may be that continued government willingness to push the applications of space technology and to bear the costs and risks of the research, development, and demonstration phases of commercializing those applications is the only way for them to become reality, at least in the short to midterm.

One area of policy and institutional controversy during the Nixon and Ford administrations was advanced communications. In 1973, NASA was ordered to end its communications R&D efforts, on the grounds that the space communications business was far enough advanced so that it should be totally a private sector responsibility. The consequence of this decision was that the U.S. private sector concentrated on only those aspects of space communications which had the promise of early commercial payoff. Other governments have provided R&D support for advanced space communications development, leading to increasing international competition with U.S. firms for sales of advanced communication satellites. This situation led the Carter administration in 1978 to decide that the potential economic and social benefits of communications satellites for both private and public sector use were not being adequately tended to by private sector R&D. The Carter administration reestablished a NASA research effort in the advanced space communications area and charged the National Telecommunications and Information Administration of the Department of Commerce with assisting in market aggregation and possible development of domestic and international public satellite communication services.

From "Preeminence" to "Leadership"

In 1961, John Kennedy committed the United States to a policy of "preeminence" in all areas of space activity. The notion that the United States should maintain a position of "leadership" in space activity has been repeated by each chief executive since Kennedy.

As other countries in Europe, Asia, and South America develop independent space capabilities and as the Soviet Union continues an extremely active space effort, the meanings for the 1980s of the terms "leadership" and "preeminence" are less than clear. One possibility is for the United States to compete with other nations across the board in all areas of space activity, from the development of large, permanent manned structures in orbit, through various types of space applications, to exploration of the cosmos. Another option is to focus U.S. space priorities in areas of high national payoff (which would include international leadership in those areas). Another option is to view application activities in space as competitors with Earth-bound enterprises, and to undertake them only when they are the most efficient means of meeting broader national objectives.

The initial impact of the commitment to across-the-board preeminence was to create in NASA an agency with the structure, institutional relationships, and organizational culture needed to carry out a high priority, nationally mobilized effort in the development of large scale technology. NASA, at least in formal terms, remains today an organization designed for such purposes, but the meaning of a national commitment to leadership in space activities is much less clear than it was during the peak of the Apollo program in the mid 1960s. As space activities have matured, and as they promise to become even more a routine part of a variety of government and private sector activities over the coming decade, a major institutional issue is whether a single central space agency with the desire and structure for carrying out an integrated, high-priority national space effort in the civilian sector is an anomaly.

The International Context: Collaboration or Competition?

During the 1960s, NASA developed international cooperative programs which were clearly secondary in priority to using space technology

as a demonstration of national technical resources. Almost all of NASA's
international activities were scientific in character* and were carried out
under policy guidelines which kept them limited in scope, including the
notions that cooperation had to be based on mutual scientific benefit and
that there would be no exchange of funds between the United States and
its partners in international space activities.[10] This limited concept of international cooperation was broadened during the 1970s to the applications area, as a number of nations became interested in the Landsat program, building their own ground stations or otherwise receiving Landsat
data, and for the first time paying NASA a fee for access to the remote-
sensing satellites. Other applications efforts had international dimensions; for example, the Applications Technology Satellite and Communications Technology Satellite programs demonstrated some of the
uses of communications satellites for education and health care in both
developing and industrialized countries.

Also during the 1970s, there was limited use of international
cooperation in space technology to serve what were explicitly foreign
policy goals. The leading example was U.S.-USSR cooperation in the
Apollo-Soyuz Test Project. Increasingly, the potential of space as a tool of
our foreign assistance program and as a means of demonstrating our concern for the developing countries has led to assistance programs related to
the utilization of remote-sensing data for a variety of third and fourth-
world countries.

During the same time period, there was the beginning of cooperation with our major industrial partners (and potential competitors) in
space technology development. The European Space Agency assumed the
responsibility for developing the Spacelab, which is to be flown on the
Space Shuttle as a base for orbital scientific experiments requiring the
presence of human experimenters. The relationships with other industrial
countries with respect to space technology are, however, somewhat am-
bivalent, because of possible economic returns on a substantial scale from
space activities and because of the desire of the United States to either
maintain or establish a competitive advantage in such areas of future
economic payoffs.

As other major nations develop advanced space technology, the mixture between international competition and international collaboration in
space should be a dynamic one. Competition between U.S. and European

---

* A major exception was the set of international agreements required to establish a global tracking
network.

launch vehicles for payloads in the 1980s is just one example. A number of issues being debated in international forums could affect U.S. civilian space activities in the coming decades. Examples are the actions of the World Administrative Radio Conferences in allocating frequencies (and potentially slots in geosynchronous orbit) and the debate in the United Nations on a Moon Treaty.

The Soviet Union, West Germany, France, Japan, Brazil—and indeed a number of other countries—are allocating significant resources to space R&D. In coming years, the U.S. civilian space program will function in a quite different international context than has been the case. The institutional implications of this changed context—for example, how to relate space activities to foreign policy objectives and how to carry out the diplomacy required to support our space objectives—require examination.

Current Space Policy Principles

This section will examine the current status of space policy from the perspective of its relation to the present institutional structure of the national space effort just described. The purpose of this examination is to identify those areas of institutional stress which will condition the ability of the United States to carry out whatever objectives for space it chooses in the 1980s and beyond.

The space policy principles of the 1957–1962 period described earlier represented a consensus arrived at after vigorous debate and under the competitive stimulus of Soviet space accomplishments. The sense of urgency that led to this consensus, which included setting a challenging goal as a central theme of the U.S. national space program, has been largely missing in the 10-year debate on appropriate principles to guide U.S. efforts in space in the post-Apollo period. That policy debate, indeed, still continues. Although some interim principles of U.S. space policy in 1980 are specified below, they do not command the kind of broad support among interested parties that the earlier set of policy principles did. A number of views on the appropriate pace and direction of U.S. space activities and of the policy principles which should underpin those activities are still represented in the policy debate.

The Carter administration articulated a U.S. space policy for the 1980s, but challenges to this policy concept have arisen from key members of both the Senate and the House, from various aerospace industry groups and representatives of the aerospace profession, and from the rapidly growing network of interest groups which focus on space policy.[11] The likely policy stance of the Reagan administration is, at the time of writing, still very unclear. Lacking any consensus on space policy, the U.S. civilian space effort is continuing largely on the momentum established by the Apollo project and the other high intensity activities of the 1960s and continued during the 1970s with the development of a new technological capability for space operations in the form of the Space Transportation System.

At issue in the current space policy debate are such questions as:

- Should long-term goals for space be articulated, or should the U.S. civilian space program be primarily an evolutionary undertaking?
- Is there a need for a commitment to a major new technological enterprise, such as the development of a permanent manned orbital facility, to serve as a focal point for the next decade in space, as Apollo did in the 1960s and the Space Shuttle in the 1970s?
- What role should men (and women) play in future activities in space?
- How aggressively should the government support the development and demonstration of potential applications of space technology to provide benefits on Earth?

A key element of the original space policy was that certain types of space activities, particularly large-scale demonstrations of technological capability, would be undertaken for what were fundamentally political motivations. This policy, as was mentioned earlier, was established by President Kennedy and was a reversal of the set of justifications for space programs accepted by the Eisenhower administration. It appears as if the United States has returned to that original set of justifications, which saw the development of space technology only as a means, not as an end in itself. The Carter administration in its space policy statement, noting that "more and more, space is becoming a place to work," suggested that *"activities will be pursued in space when it appears that national objectives can most efficiently be met through space activities."* [12]

This policy principle is applicable most directly to the economic, social, and military applications of space technology. It recognizes the rapidly maturing state of space capabilities and suggests that space programs are increasingly recognized as means to some desirable end, not ends in themselves. Not only does current policy reject the notion of space as an arena for symbolic political competition, but it also indicates that there may be limits on the investment of resources in space activities aimed at scientific returns. The same space policy statement, while emphasizing U.S. commitment to a space science and exploration program which "retains the challenge and excitement" of new discoveries, also notes the need for "short-term flexibility to impose fiscal constraints" when necessary. The combination of *a priori* requirements for cost-effectiveness and the recognition that general budget constraints are important determinants of the level of government investment in space activities underpin a much more limited concept of the importance of space activities on the national agenda than was the case under the space policy of 1961.

It should be noted that the concept of a "lowered profile" for the U.S. space program did not originate with the presidency of Jimmy Carter. The Carter space policy was to a large degree, a continuation of that adopted during the immediate post-Apollo period by Richard Nixon, who noted in 1970 that "what we do in space from here on must become a normal and regular part of our national life and must therefore be planned in conjunction with all of the other undertakings which are also important to us." [13] In 1972, the Nixon administration did make a commitment to the Space Shuttle, a major technology development program, but that decision, to a large degree, was made without relating it to any overriding sense of policy objectives; there was a generalized notion that a less expensive and more flexible capability for routine space operations was likely to be a rewarding investment of national resources.[14] The Shuttle decision had few parallels with the decision to go to the Moon a decade earlier; it was a commitment to technological development without a clear link to an overriding political or other policy justification. The Carter administration rejected an Apollo-like commitment to another major space technology project, suggesting that "it is neither feasible nor necessary at this time to commit the United States to a high-challenge space engineering initiative comparable to Apollo."[15]

The earlier space policy of the United States stressed preeminence, particularly in its implementation by large scale technological enterprises, as an overriding policy goal. This principle has been replaced by one which stresses *balance among scientific exploration, applications of space technology, and technology development.* Within this balanced strategy there is an emphasis on Earth-oriented applications of space technology, whether they be social, economic, or military in nature. This emphasis on balance among various types of space activities is also one that stems from earlier administrations. In the same 1970 statement mentioned above, Richard Nixon had noted "many critical problems here on this planet make high priority demands on our attention and resources. By no means should we allow our space program to stagnate. But—with the entire future and the entire universe before us—we should not try to do everything at once. Our approach to space must be bold—but it must also be balanced."[16]

More specifically, the United States has given increased priority over the past decade to demonstrated and potential military applications of space technology. A "growth sector" over the past decade has been research, development, demonstration, and operation of space-based military systems for carrying out essential military support functions such as communications, command, and control; early warning; strategic surveillance; navigation; and weather forecasting. An expanded list of military applications in space is now under consideration and may be more likely to gain political and budgetary support than any of the contending applications of space technology for civilian purposes.

One principle of U.S. space policy established in the late 1950s has remained valid in the current situation. That principle is that *civilian, military, and intelligence space activities will be carried out in separate institutional structures.* A recent presidential review confirmed the current management relations in the government's space effort; and thus NASA, DoD, the intelligence community, and NOAA each remain responsible for different parts of the government space program. However, with the maturing of space technology developed under these various programs and with the emphasis on increased efficiency and resource conservation, there is *more emphasis than before on transfer of technology among the various government space programs and on jointly-funded and jointly-managed programs serving multiple objectives.*

The emphasis on technology-sharing and joint programs will place increased demands on mechanisms for program as well as policy coordination. Because it is in the nature of most large-scale bureaucratic organizations to resist sharing resources and to prefer individually managed programs, and because military and intelligence programs can "hide" technology behind security classifications, the kind of presidential and congressional pressure now being exerted on the national space effort to support the idea of resource-sharing is probably necessary, if the twin principles of maintaining the separation between programs and attempting to carry out truly national efforts are to be successful.

Another policy principle stemming from the beginning of the U.S. space program which remains unaltered is that *NASA is limited to research and development activities only and will not operate space systems.* * NASA's role as an R&D-only agency was revalidated during the

---

* As mentioned earlier, an exception to this principle is NASA's operational role as a provider of launch services. This role is likely to be reexamined as the Space Shuttle reaches routine operational status.

consideration of national policy on remote sensing in 1979. Among others, the NASA leadership believed that the agency could best continue to make a contribution to the national space program by restricting itself to R&D activities. A consequence of this policy principle in a period in which various applications of space technology, particularly in the land and ocean observation areas, approach operational status is that some other entity, either public or private, must be assigned responsibility for the operation of space applications systems. Currently, the responsibility within government for Earth observation from space has been assigned to a single agency, NOAA, rather than spreading it among several federal agencies or creating a new government agency with specific responsibilities for Earth observations. In coming years, NOAA may well become as much of a space agency as NASA is today, even though NASA will continue to do the research leading towards operational space applications, including related ground segments, and will continue its role as the agency in charge of space science and exploration.

Another policy principle which has remained unchanged in general form, but rather different in operational meaning, is that the *government will actively encourage private sector involvement in the uses of space technology, while also sponsoring research in areas of potential commercial application.* The development of relationships between public sector and private sector interests in space applications has proved a particularly difficult task. The transfer of the results of government-funded research on communications satellite technology to application in privately-owned, operational, communications satellite systems was straightforward in comparison to arranging for private sector involvement in areas such as navigation* and, particularly, remote sensing. With civilian space activities within the government now divided between NASA, NOAA, and a number of other federal agencies, relationships between the private sector and government space programs are even more complex. Private sector involvement with NASA in the design of research efforts in space applications is likely to continue to be necessary, as will be relationships between NOAA and private sector entities interested in the commercial potential of Earth observation systems.

Finally, the international dimensions of space activity are receiving considerably more attention at the present time than had been the case earlier. Congress has been particularly interested in international cooperation in space activities. Because other industrial countries are developing

---

* Most of the work leading to space-based navigation systems has been carried out by DoD, and making that capability available for civilian applications is proving problematical. NASA has undertaken only limited work related to space-based navigation or position-location systems.

substantial civilian space programs emphasizing applications of space technology, the United States finds itself in a situation in which *opportunities for meaningful cooperation in space are mixed with the potential for significant competition in areas of high economic and social payoff.* Also, other nations, perhaps more than the United States, still undertake space programs as means of enhancing national prestige, and this motivation constrains cooperative efforts. No clear policy principle relating to the international aspects of U.S. space activities has yet emerged from space policy debate of the last decade; this is an area of policy development which is "ripe" for increased attention.

Concluding Comments

As a new stage in the evolution of U.S. space activity is entered with the imminent launch of the Space Shuttle, a meeting such as this—aimed at focusing the attention of historical professionals on opportunities for study presented by space programs—seems to me to be quite appropriate. The space program deserves the attention of academic historians and their students, because academia provides the unconstrained and broad-gauged context within which it can be best understood. Future generations are almost certain to view mankind's first tentative expeditions away from its home planet as major historical events. From that perspective, it is a privilege to be in at the beginning.

Source Notes

1. The following account of the origins and early evolution of U.S. space policy and institutions is drawn from a number of sources, including: John Logsdon, *The Decision to go to the Moon*, (Cambridge, Mass.: MIT Press, 1970); Arthur L. Levine, *The Future of the U.S. Space Program*, (New York: Praeger Publishers, 1975); Robert L. Rosholt, *An Administrative History of NASA, 1958-1963*, (Washington, D.C.: NASA, 1966); and Enid Bok Schoettle, "The Establishment of NASA", in *Knowledge of Power: Essays on Science and Government*, ed. by Sanford A. Lakoff (New York: The Free Press, 1966).
2. See Logsdon, *op. cit.*, for a full account of how the Apollo decision was made and how it represented a fundamental reversal of previous space policy.
3. Mose L. Harvey in "Preeminence in Space: Still a Critical National Issue," *Orbis*, Vol. XII, No. 4 (Winter 1969), defines preeminence as "achieving broadly based capabilities with regard to all aspects of the space environment and then constantly building upon and adding to those capabilities." (p. 959). See also Vernon Van Dyke, *Pride and Power* (Urbana: University of Illinois Press, 1964) for an analysis of the motivations underpinning early U.S. space policy.
4. For an account of the development of U.S. policy on satellite communications, see Jonathan F. Galloway, *The Politics and Technology of Satellite Communications* (Lexington, Mass.: D.C. Heath and Co., 1972).
5. The works by Levine, Rosholt, Schoettle, and Logsdon cited earlier do contain such an account.
6. The only public announcement of the results of this review was in the form of a June 20, 1978 press release from the White House.
7. Quoted in Levine, p. 66.
8. For a detailed account of the NASA/Weather Bureau dispute, see Richard Chapman, *TIROS-NIMBUS: Administrative, Political, and Technological Problems of Developing U.S. Weather Satellites* (Syracuse, N.Y.: Interuniversity Case Program, Inc., 1972).
9. See Galloway, *op. cit.*, for a full discussion of this debate.
10. The foundations of U.S. policy toward international cooperation are described by Arnold Frutkin, *International Space Cooperation* (Englewood Cliffs, N.J.: Prentice Hall, 1965) and criticized by Don Kash, *The Politics of Space Cooperation* (West Lafayette, Ind: Purdue University Studies, 1967); there is no later analytic treatment of international space cooperation.
11. It is impossible to summarize the positions held by all parties in this debate, both in terms of overall space objectives and in terms of the priority to be given to specific programs or program areas. In Congress, Senators Adlai Stevenson and Harrison Schmitt and Representative George Brown have each proposed bills establishing national space policy, and organizations as diverse as the American Institute of Aeronautics and Astronautics, the L-5 Society, the Planetary Society, Gerald O'Neill's Space Studies Institute, and the National Space Institute, among others, have proposed different approaches to the national space effort. In addition, NASA through various advisory councils, summer studies, and in-house planning is seeking to define both technological possibilities and appropriate goals for the future.
12. "White House Fact Sheet on U.S. Civil Space Policy," October 11, 1978.
13. Statement by the President, March 7, 1970.
14. For an account of the Space Shuttle decision, see John M. Logsdon, "The Space Shuttle Decision: Technology and Political Choice," *Journal of Contemporary Business*, Vol. 7, No. 3 (Winter 1979), pp. 13–30.
15. "White House Fact Sheet on U.S. Civil Space Policy," October 11, 1978.
16. Statement by the President, March 7, 1970.

SPACE-AGE EUROPE, 1957–1980
Walter A. McDougall

"Europe will be made in space . . . or not at all!"
Orio Giarini

"Il ne faut pas espérer pour entreprendre ni reussir pour persévérer."
William the Silent

Soon after the Soviet *Sputnik* opened the frontier of outer space, European scientists, industrialists, and politicans began to clamor for rapid entry into the space age by Europe, the cradle of modern technology. It took 22 years before the European Space Agency (ESA), on Christmas Eve 1979, finally achieved successful orbit of a European-designed spacecraft riding on a European booster, the Ariane, from its equatorial spaceport in French Guiana. The launch was beamed live (via the American-built Intelsat IV communications satellite) to French television. But the viewers—and the newsmen themselves—were so unused to such affairs that each time the countdown went on another "hold" they reacted hysterically as if the whole program were about to be cancelled. This calls to mind another anecdote from a friend who watched the coverage of the first Moon landing in 1969 in the company of a peasant family in the South of France. They were curiously blasé about the whole affair—until the report that President Nixon was about to converse with the astronauts on the Moon. Madame excitedly called the family to watch: "Look! The President of the United States, he is going to telephone the Moon . . . and we cannot even get a line to Paris!"

In these vignettes are illustrated essential themes in the first chapter of space-age Europe: tardy and hesitant enthusiasm, a certain naiveté, and public apathy to events that do not impinge on quotidian reality. In tired Europe, the age of adventure sometimes seems closed, but it is perhaps enough that there is a European chapter in space at all. In fact, the response of the major states to the challenges of *Sputnik* and *Apollo* reflect their very adjustment to the postwar world itself, a world in which the old continent struggles to find its proper place amidst superpower hegemony, decolonization, welfare statism, fitful integration, and, above all, perpetual technological revolution.

The first European implications of *Sputnik* were military. Now that the Soviets demonstrated an intercontinental ballistic missile capability to threaten the American homeland, was the U.S. nuclear deterrent still credible? Would America risk New York or Chicago to save Berlin or

Paris? And if not, could second-rank powers like Britain or France see to their own defense? Only six months after *Sputnik*, Charles DeGaulle was called out of retirement to lead a nation smarting from Dien Bien Phu, Suez, and Algeria. His *"certaine idée"* of a glorious France rested not only on rhetoric, but on a vision of technological self-sufficiency in defense and industry. In five years, French R&D spending increased four-fold, yielding a vigorous nuclear power program, an independent strategic deterrent, and the world's space program. Benefiting from its country's military missile research, the French space agency "cut" a series of precious stones— rockets called the Agate, Topaze, Rubis—until in 1965, a Diamant launcher lifted a French satellite into orbit from the Sahara desert test range. There also followed the deployment of land- and submarine-based missiles, the *Force de Frappe,* and in our own day, the beginnings of a military space program.

The French could not hope to match the space and missile efforts of the U.S. and USSR. But that was never their intent. Militarily, the French relied on the crude "city-busting" deterrence of the mutual-assured-destruction doctrine. In terms of general technology, they envisioned a world of multipolar competition in which Europe would evolve away from both Cold War camps. What was important, therefore, was that France assure herself the position of first among European equals. The French space program would help to establish French primacy in the European community.

The British, on the other hand, reacted to *Sputnik* by throwing in the towel. Their V-bomber force would soon be obsolete, but they abandoned their missile effort and resigned themselves to dependence on their "special relationship" with the U.S.—the relationship that DeGaulle so despised. But lest their first-generation intermediate-range ballistic missile go to waste, the British offered the rocket, the Blue Streak, to Europe as a whole, to serve as the first stage of a European space booster. Meanwhile, an international committee of scientists organized by Pierre Auger lobbied governments on behalf of a space science program. From these two early initiatives the European space program emerged, dedicated to admitting European science and industry to this latest and most exciting human enterprise.

It seemed like a good idea at the time. France, Italy, West Germany, and the Benelux countries had just formed the Common Market and EURATOM. A cooperative space effort was a logical step. Morever, the vast expense involved suggested the pooling of resources. So in the early 1960s, the European Space Research Organization (ESRO) and the European Launch Development Organization (ELDO) were born. The two agencies became embarrassing examples of how *not* to generate high technology.

ESRO's member countries* proposed to design payloads for satellites to be launched by NASA and eventually by ELDO. But thanks to organizational problems, inexperience, and underfunding, it was not until 1967 that the experimental ESRO 1 was in orbit. By that time Britain and Italy were already pleading straitened finances while all member governments were goading ESRO to deemphasize science in favor of commercial applications satellites with benefits perceptible to parliaments and publics. ESRO founded some impressive facilities in its early years, e.g., the spacecraft design laboratory at Noordwijk, Netherlands; a European space operations center in Darmstadt, West Germany; ground stations in Spain, Belgium, and Italy; and a sounding rocket range in Kiruna, Sweden—but there were endless startup problems associated with them. Discord also stemmed from disproportional distribution of contracts to the member states, the problem of *juste retour*. France, for instance, received a percentage of ESRO contracts twice the level of her contributions, and less favored nations complained that such practice only perpetuated their industrial inferiority. This pointed up a grievous problem with cooperative R&D: efficiency demands that contracts go to the most qualified bidder, but politics demand "affirmative action" for less experienced firms in countries hoping to play "technological catch-up." Either the poor help to subsidize the rich, or the rich subsidize mediocrity in the short run and new competition in the long run.

While ESRO struggled, ELDO fizzled. It had projected a European booster consisting of the Blue Streak as first stage, a French-built second stage, a German third (or apogee) stage, and an Italian test satellite. Anyone familiar with the difficulties of systems interface in the American program can imagine the boondoggle of an international rocket. By 1969, the Europa booster had gone through numerous design changes, had never flown, and was 350 percent over initial budget. Veterans of those days have written positively impolite accounts of their experiences with foreign colleagues. One of the more tolerant was this depiction of national temperaments: "Whenever we faced a technical or administrative problem requiring improvisation, the French would stubbornly refuse to violate any hard-won principle of procedure; the Germans would endorse

---

* Belgium, Denmark, France, Germany (West), Italy, Netherlands, Spain, Sweden, Switzerland, and the United Kingdom. Austria and Norway had observer status.

the principle, then list all conceivable exceptions; the Italians would excitedly urge re-negotiation of the principle to accomodate the offending contingency, while the British would cheerfully accept any improvisation without question—so long as under no circumstances would it serve as a precedent!'' [1] Others complain that European ministries used ESRO and ELDO as dumping grounds for deadwood personnel. In any case, the babble of tongues only exacerbated the habitual lack of communication among scientists, engineers, and bureaucrats.

By the late 1960s, the European space effort was a shambles. That it persisted was due in part to a second shock wave from abroad—the first had been the Soviet *Sputnik*, the second was America's vigorous reaction to *Sputnik*. From aboard, America's heady expansion of the 1960s seemed to comprise nothing less than a second industrial "takeoff," illustrated by her space triumphs, booming economic growth, and ubiquitous foreign investment. It all seemed to stem from what one French economist called "the keys of power": government forcefeeding of science, technology, education, and investment in "point sectors" of the economy, especially aerospace. Americans themselves may never have felt entirely comfortable with the massive increase in state stimulation of economic and social change, but the American model made a profound impression on a Europe already inclined toward *étatisme*. European economists and pundits swallowed the arguments of the Kennedy and Johnson administrations on behalf of big-government R&D even more than we did ourselves. The visionary Apollo program and its technological and managerial "fall-out" had seemed to open a vast technology gap between the U.S. and Europe. Talented Europeans fled to the advanced laboratories of America, causing a "brain drain" that further handicapped European science. It seemed the old industrial and imperial powers would face a future of "industrial helotry" if Europe did not match the technological surge of the U.S. DeGaulle himself intoned: "We must invest constantly, push relentlessly our technology and scientific research to avoid sinking into a bitter mediocrity and being colonized by the invention and capacity of other nations." [2]

For European business the apparent threat from America, later popularized by Jean-Jacques Servan-Schreiber's *Le Défi américain*, was the best propaganda for higher space budgets. As early as 1961, European industrialists had formed a private lobby called EUROSPACE. Throughout the 1960s it beat the drum for state-financed R&D, warning Europeans against their tendency to sniff at the technical accomplishments of boorish Americans while taking comfort in their superior culture. "Carthage was a flourishing culture," observed the president of EUROSPACE, "when it met its doom. And it was not the exceptional culture or eloquence of Rome that allowed her in turn to resist the pressure of barbarians." Rather, "the evolution of all humanity is closely linked to technological progress. . . . If Europe does not regain her place in the first rank of technological civilization it will soon be too late."[3] The Germans expressed this as *Torschlusspanik:* Europe must leap now or the door to the space age would slam shut. The Italian government called for a "technological Marshall Plan." In Britain, Harold Wilson proposed a "European Technological Community."

These fears and exhortations of the late 1960s proved to be exaggerated. But they seemed to be confirmed at the time by the one profit-making enterprise in space applications—Intelsat. This consortium for international telecommunications satellites founded by 19 nations in 1964 was an American show. The U.S. controlled 61 percent of the voting authority and all the technology. It was even managed under contract by the U.S. Communications Satellite Corporation, which was dominated in turn by such giants as American Telephone and Telegraph (AT&T). This situation irked the Europeans, but there was no competing with the Americans since U.S. export laws forbade sale of launch technology to Europe, and NASA was under orders not to provide launch service for satellites able to compete with Intelsat. Here was precisely the sort of dependency of which the French always warned.

The early 1970s were consequently a confused time of negotiation and reorganization for the extant and aspiring space powers. Apollo was winding down and the Shuttle being planned. The U.S. invited the Europeans to cooperate more closely in space while talking compromise on Intelsat and satellite launch policy. Why should Europe waste millions to duplicate American efforts? This was persuasive, but on the other side the French continued to campaign for independence, offering to take the lead in a reinvigorated European effort. The result was a grand compromise. In 1975, a new European Space Agency absorbed ESRO and ELDO, drawing on their facilities and experience, but dedicated to avoiding their shortcomings. A new system of *à la carte* financing, by which members need pay for only the programs they support, and centralized management of major programs under a single country, promised both *juste retour* and improved efficiency. European aerospace firms also promoted equitable subcontracting through formation of private international consortia.

ESA was built around three main projects, all now nearing completion, which reflected the compromise between independence and collaboration with the U.S. To Britain went the major role in funding and developing the MARECS marine navigation satellite system; West Germany received major responsibility for the sophisticated Spacelab,* a space sciences module custom-made for the cargo bay of the U.S. Shuttle. Finally, France charged ahead with development of Ariane, a heavy satellite launcher capable of boosting communications satellites into high geosynchronous orbits. Meanwhile, the U.S. relinquished control of Intelsat in a new, permanent convention—and European and Third World delegates promptly voted to deny a launch contract to the U.S. and sign on with Ariane.

It would appear at present that Europe has finally succeeded in fashioning the diplomatic, organizational, and technical prerequisites for a sustained, effective space program. European aerospace and electronics firms—often bearing worthy risks in light of fickle government policies and uncertain markets—have reached state-of-the-art expertise in chosen fields. But the future of Europe in space is still far from assured. ESA is still troubled by political and economic difficulties, and the central goals of European space activity are still unenunciated after 20 years. Both Eurospace and ESA's Director-General, E. Quistgaard of Sweden, pressed again in 1981 for a plan of space development for the decade of the 1980s. As in the past, member governments refused to look beyond immediate budgetary cycles or enunciate long-range goals. Funding should continue at current levels of about $840 million per year, enough to support an approved second launch pad at Kourou, French Guiana, development of the improved Ariane 2 and 3, and possibly an experimental Earth resources satellite. But new starts are few, and scientific missions like Giotto, the gripping rendezvous with Halley's Comet, are small potatoes. In fairness, one must recognize the inability of the U.S., freed of multilateral confusion, to draft long-term plans of its own. But as Quistgaard laments, all the problems of the individual European governments *and* of the balked process of integration weigh upon those charged with getting Europe into space.

Every member state contributes unique strengths and weaknesses to ESA. But the character of the European space program from its inception has been shaped above all by France. ESA still lies in the shadow of a Gaullist Europe that never happened. Britain never could have led Europe into space. Her tired taxpayers and confused bureaucrats were

---

* The prime contractor for Spacelab was the German firm ERNO, a subsidiary of VFW. Its development cost was $800 million. The first operational Spacelab mission, featuring a German astronaut, was scheduled to ride the Shuttle in late 1983.

most skeptical of glamorous R&D, had no defense motive, and were of two minds about European integration. Germany was the founder of modern rocketry, but she was barred from missile R&D because of the unpleasant use she made of the V-2. Only France was capable of a gritty national effort and of taking the lead in cooperative programs. And the advent of DeGaulle by historical accident in 1958 meant that France's mission in Europe, and Europe's in the world, were defined in terms exceptionally favorable to space activities. But it also meant that Europe in space would be stamped with Gaullism. ELDO and ESRO—instead of helping to forge a united Europe—served instead to elevate France within a Europe in which national prerogatives would be closely guarded and international institutions promoted mostly as a tool against the Anglo-Saxons.

France dominated ESRO and ELDO, and her industries benefited most from them. France's cooperation policies with Europe, NASA, or the Soviet Union were designed as much to tap foreign funds and skills for the benefit of her own national program as the other way around. It was France that led the campaign against dependence on America, even when logic may have dictated a division of labor. It was France that bartered her indispensable cooperation for ESA's approval of a Franco-European launcher and Franco-European communications satellite program. And it is France that benefits most today from the prestige, technology, and military applications of European space research.

This is not to say that France has exploited others. She has consistently made the largest contributions to European space funds, currently 25 percent. Nor is it to say that France's partners in ESA do not glean rewards commensurate with their participation. Nor is it even clear that the Gaullist insistence on French independence was not farsighted, given the uncertainties of world politics and power balances over the long run. But the fact remains that French space policy has been doggedly nationalistic, and that the European space establishment—as are all other European institutions—is a hostage to that policy.

What of domestic support for space activity? Here again, the role of Gaullism is critical. To be sure, public opinion has had its cycles, as the U.S. European excitement and worry about technological inferiority peaked around 1968, and by the early 1970s, Europeans, too, were becoming disenchanted with technology as a social panacea. Thus, even as ESA came into being, European opinion was cautious on space spending. ESA and member governments have sometimes been uncertain what posture is best for the protection of space budgets: proud publicity or a low profile. Today the chances are good that the man on the street in

Lyon, not to mention Naples or Liverpool, is scarcely aware of ESA or Ariane. But current apathy ought not to obscure the deep domestic significance of the space effort. For the legitimacy of a French or European thrust into the cosmos is rooted in the historical circumstances of its birth, in the role that technology was to play in the stabilization of the Fifth Republic. DeGaulle declared himself a defender of traditional France in social relations, politics, and culture, even as he decreed the end of imperial France (with retreat from Algeria), the end of European France (with resistance to further integration), the end of atlanticist France (with withdrawal from NATO), and the end of socialist France (with defeat of the left). In order to preserve tradition in the abstract realms of French life, DeGaulle proposed to overthrow tradition in the material realm. Technological revolution would translate abroad into the prestige and independence of French tradition, and at home in the seductive vision of the future that invited France and Europe to imagine themselves "in the year 2000," that inescapable slogan of contemporary Europe.

Hence the legitimacy of a Gaullist regime that claimed to play midwife to the future even as it invoked the past. What DeGaulle actually offered was a French version of our own "Republic of Technology," in which social and international challenges alike are spirited away (in theory) through the genie of the technological fix; where leaders pose as defenders of tradition even as they undermine it indirectly through technological revolution. In a Europe that is frankly nonideological, materialistic, and atheistic, this pattern of technetronic [4] politics is discernible not only in France, but everywhere.

Has high-technology investment really transformed Europe? This is a tough question, given the difficulties of measuring second-order consequences of R&D. European industry has certainly escaped "backwater" status, and western Europe is again part of the world technological vanguard. But the effect of space activity on Europe must still be sought in the political, not economic, realm. For the Europeans chose to reject a global division of labor in space, and thus to duplicate many U.S. and Soviet achievements. And for what? Arianespace, the new commercial firm, may show a profit, but only because its R&D costs were absorbed by European taxpayers and because its launch price may be subsidized to compete with the Shuttle. In any case, Ariane only matches a capability the U.S. had had for two decades. As for the goal of industrial prowess, European motives were again largely political, as demonstrated by the fact that European aerospace firms have become semi-public "chartered companies" of the state. The recent German union of MBB (Messerschmidt) and VFW is only the latest in a series of forced mergers that previously produced British Aerospace, France's Aerospatiale, and Italy's Aerospaziale—all for the purpose of competing, not in capitalistic, but in mercantilistic fashion, with the giant American firms and with each other, in a business otherwise too big for "little" Europe.

As the 1980s mature, it is entirely possible that even the concentration of resources within each European state, even the pooling of resources among European states may not suffice to sustain an independent European role in space without sharply higher levels of spending, which in turn may prove politically impossible. Even at the two peaks of the mid-1960s and late 1970s, Europe spent only a driblet on space: 0.1 percent of combined GNP versus 1.5 percent for the USSR and between 0.5 and 1.0 percent for the U.S. In per capita terms, the superpowers have spent 20 times more than Europe. As the U.S. now gears up for another space/defense push, and as reusable spacecraft, antisatellite weapons, and permanent space stations emerge as near-term prospects, the future of a coherent, independent European space effort is dubious. By around 1985, with Ariane and Spacelab and MARECS completed, the Europeans will again have to face the question "L'espace pour quoi faire?" Member governments may have to:

• Ante up a considerable investment on a truly multilateral basis, implying unprecedented political unity;
• Continue such programs as Ariane permits, but otherwise accept a role of "subcontractor" to the U.S. in the many fields of space exploitation made possible by the Shuttle;
• Throw in the towel, cutting back state expenditures on space and accepting a reduced or very different view of the role of western European states in the world.

Severe economic crisis could force the third course. Otherwise, the French will remain independent and ambitious. The Americans will continue to extend the hand of cooperation, in part to relieve their own budgetary strains. The Germans, whose wealth and expertise are attractive, will be in the middle, wooed by Washington and Paris as they were in DeGaulle's day. For the Shuttle may open up a universe of possibilities in space industrialization, weaponry, satellite repair and recovery, permanent manned stations, and more. The Germans in turn will be enticed—and the irony may come to pass that decisions made in Bonn and not Paris will finally determine what "Europe in the year 2000" will be doing in outer space. Giarini's intuition may soon prove valid, that "Europe will be made in space . . . or not at all."

Source Notes

1. Jacques Tassin, *Vers l'Europe spatiale* (Paris, 1970). pp. 98–99, a somewhat embellished paraphrase.
2. Charles De Gaulle, *Addresses to the French Nation*, 1964.
3. Jean Delorme in EUROSPACE, *Europe and Space: An Assessment and Prospects* (Konstanz, 1971), pp. 6ff.
4. The neologism is Zbigniew Brzezinski's. See *Between Two Ages: America's Role in the Technetronic Era* (New York, 1970).

References

For general accounts of the world's space programs:

House Committee on Science and Technology, *World Wide Space Activities*, 90th Congress, (Washington, 1977).

Alain Dupas, *La lutte pour l'espace* (Paris, 1977).

On European space efforts:

Orio Giarini, *L'Europe et l'espace* (Lausanne, 1968).

Robert Gilpin, *France in the Age of the Scientific State*, (Princeton, 1968).

Jean-Jacques Servan-Schreiber, *The American Challenge*, (New York, 1968).

Jacques Tassin, *Vers l'Europe spatiale*, (Paris, 1970).

Georges Thomson, *La politique spatiale de l'Europe* (Dijon, 1976).

Norman Vig, *Science and Technology in British Politics*, (Oxford, 1968).

On the Soviet space program:

Nicholas Daniloff, *The Kremlin and the Cosmos* (New York, 1972).

James Oberg, *Red Star in Orbit* (New York, 1980).

Charles Sheldon, *U.S. and Soviet Progress in Space: Summary Data through 1973 and a Forward Look* (Washington, 1974).

Leonid Vladimirov, *The Russian Space Bluff* (London, 1971).

On Intelsat:

Jonathon F. Galloway, *The Politics and Technology of Satellite Communications* (Lexington, Mass., 1972).

SPACE ACTIVITIES IN THE SOVIET UNION, JAPAN, AND THE PEOPLE'S REPUBLIC OF CHINA*

Edward C. Ezell

The launch of *Sputnik 1* by the Soviet Union on October 4, 1957, began the era of modern spaceflight. Within four months, the United States had joined the "space club" with the successful orbiting of *Explorer 1*. Seven and a half years passed before a third nation joined this exclusive association; France put its *A1* satellite into orbit on November 26, 1965. Japan and the People's Republic of China became Asia's representatives in space in 1970; the Japanese *Osumi* and the Chinese *East Is Red* were orbited on February 11 and April 24, respectively. The final member of the "Space Six," the United Kingdom, launched the satellite *Prospero* on October 28, 1971. Comparative data for these satellite launches are given in table 1. In 1981, the European Space Agency will likely become the seventh organization to boost its own payload into orbit. As the number of spacefaring nations grows, we should look back and examine what common and divergent motivations have sparked this thrust into space. For the purposes of the Yale University National Aeronautics and Space Administration (NASA) Conference on the History of Space Activity, this paper will concentrate on the space programs of three of the six nations that have undertaken their own space programs—the Soviet Union, Japan, and the People's Republic of China—examining briefly the types of launch vehicles they have used and the classes of spacecraft they have launched.

Motivations

To understand why six countries have engaged in such a costly enterprise as spaceflight, we must realize that for each country there existed a complex set of motivations for taking that first step. For the purposes of analysis, these motivations can be broken down into three basic categories—political and military, scientific, and practical. These categories are certainly not exclusive compartments, and I have not attempted to rate one country's justifications for undertaking a space program as more socially acceptable than another's.

It would seem that most nations made the commitment to space

---

* The opinions expressed in this paper represent those of the author and in no way should be interpreted as an official expression of the National Aeronautics and Space Administration.

once their public leaders came to see it as an acceptable and valuable activity within the context of domestic and international politics and then approved the expenditure of public funds necessary to support the venture. However, the first artificial satellite projects grew out of scientific proposals made for the International Geophysical Year (IGY) of 1957–1958, a multinational effort to study the entire planet. Several participants believed that the IGY would be enhanced by using satellites to gather geophysical and astrophysical data from above the atmosphere, and only two nations had the wealth and technology to answer the challenge of spaceflight at this early stage, the United States and the Soviet Union. The scientists involved in the IGY knew that more than scientific riches would come from the first successful flight of a manmade moon; political and psychological prestige with military overtones would be the extra bonus.

Competition between the Soviet Union and the United States for international prestige was an extension of Cold War attitudes that had existed between them since the immediate postwar years. Their alliance to defeat the Axis powers in World War II had been, in many ways, an uneasy one, and with victory over the common enemy, they had begun to view each other with increasing apprehension and mistrust. In the resultant rivalry, technology as translated into industrial capacity and military hardware became a major indicator of national prestige and power. The Soviets and Americans had emerged as victors from the World War in part because the industrial sectors of their respective societies had provided their troops in the field with the machines of war in quantities that Germany industry could not match. Among this hardware were two new weapons that would become critical in the postwar world. One was the atomic bomb developed by the United States; the other was the V-2 rocket created by Germany. The significance of the first atomic weapons was immediately apparent after Hiroshima and Nagasaki, but the promise—or threat—of ballistic rockets was seen less clearly, perhaps because the V-2 had been a less than perfect weapon. But the Soviet and American military establishments wasted no time in developing this new technology in the decade following the war, and both countries put military rockets and nuclear research on their high-priority lists.

The results of this postwar competition between the Americans and the Soviets are well known. The Soviets were the first to orbit a satellite, which was damaging enough to America's national ego; but more worrisome, they did it with an intercontinental ballistic missile that could be used to deliver a decidedly more lethal payload. The Soviets had obtained a highly visible and indisputable technological first. Americans not only

Table 1. Comparative Data for the First Satellites Launched by the Soviet Union, the United States, France, Japan, the People's Republic of China, and the United Kingdom

Date of Launch	Country (launch vehicle)	Name of Satellite (international designation)	Weight kg (lbs)	Apogee km (statute miles)	Perigee km (statute miles)	Period minutes	Inclination degrees
October 4, 1957	Soviet Union (Raketanosytel "Sputnik")	Sputnik 1 (1957-Alpha)	83.6 (184.3)	947 (588)	228 (142)	96.2	65.1
January 31, 1958	United States (Jupiter-C)	Explorer 1 (1958-Alpha)	14 (30.8)	2,530 (1,573)	360 (224)	114.8	33.3
November 26, 1965	France (Diamant)	A1 (1965-96A)	42 (92.5)	1,767 (1,098)	525 (326)	108	53
February 11, 1970	Japan (Lambda 4S)	Osumi (1970-11A)	38 (84)	5,136 (3,191.4)	525 (326)	116.1	31.4
April 24, 1970	People's Republic of China (Long March 1)	East is Red (1970-34A)	172.8 (381)	2,387 (1,483.2)	439 (272.8)	114	68.4
October 28, 1971	United Kingdom (Black Arrow)	Prospero (1971-93A)	65.8 (145)	1,540 (957)	552 (343)	106.4	82

perceived the challenge of this accomplishment but also saw it as a threat to their security and their place as the world's leading military power. As the Soviets reaped political, military, and scientific returns from their new star, American leaders embarked upon a period of deep, worried self-examination. The obvious response to the Soviet feat was an intensification of the American programs to launch a satellite and an increase in the tempo of military rocket research. Declared or not, a bilateral technological competition had begun in this new arena. The "space race" of the 1960s, at least for the United States, also became a visible civilian—and peaceful—surrogate for the more secret military arms race. It has been argued that NASA's Apollo program could be interpreted as America's way of telling the Soviet Union and the world that it was still a technological giant to reckon with. "If we can land a man on the moon, . . ."—would-be adversaries were invited to complete the sentence. The message was clear: The sophisticated technology applied to the lunar exploration project could be easily translated to military systems.

The French, under the leadership of General Charles DeGaulle, clearly understood this fact of life. Caught between the Scylla and Charybdis of Soviet and American nuclear armament, DeGaulle was convinced that the French must develop a nuclear military capability independent of the two superpowers if they hoped to maintain credibility as a military and political power. The French began development of their Diamant (Diamond) launch vehicle in the early 1960s as a nuclear weapons delivery system. Taking advantage of the first test launch of the three-stage missile, the French also orbited their first satellite on November 26, 1965 (with NASA launching another French-made satellite, the *FR1*, a few days later.) Because it had no scientific mission and carried only limited radio instrumentation, the *A1* satellite was criticized by the world's scientific community, but French military authorities readily admitted that the primary objective for the mission had been to rest the missile. Here was proof that the French nuclear *force de frappe* was indeed genuine. The French could also play the game of surrogate technology.

Japan became the fourth nation to develop the technology necessary to join the space club, but unlike the Soviets, Americans, and French, the Japanese did not use a modified military launch vehicle. Their postwar constitution forbade the construction of such offensive military hardware, allowing them only defensive military equipment. Civilian organizations interested in the scientific and practical utilization of space served as the catalysts in Japan for the development of launchers and satellites. While not as technologically advanced as the Soviets or the Americans and still not economically recovered from the Second World War, the Japanese had shared the interests of the world powers in space exploration since the IGY period. Through the Institute of Industrial Science at Tokyo University, Japan participated in the International Geophysical Year in 1958 by launching small sounding rockets capable of taking various measurements in the upper atmosphere and went on to launch successively more powerful sounding rockets in 1961, 1965, and 1966. On February 11, 1970, the Institute of Space and Aeronautical Science (formed from the merger of the Institute of Industrial Science and the Tokyo University Aeronautical Laboratory) orbited its first satellite. Japan's Lambda 4S launch vehicle was domestically developed, as was its successor M-rocket. The N-rocket launcher is a hybrid made from the McDonnell-Douglas-manufactured Delta (Thor) booster and an upper stage developed in Japan with technical assistance from Rockwell International. Mitsubishi Heavy Industries serves as the National Space Development Agency's prime contractor. The Japanese satellite program is divided between so-called practical and scientific projects; the former are conducted by the National Space Development Agency, the latter by the Institute of Space and Aeronautical Science.

Two and a half months after the Japanese launched their first satellite, military and space specialists of the People's Republic of China launched theirs. It was called *East is Red*, because it broadcast that revolutionary anthem as it orbited the Earth every 114 minutes. As had the Soviets, Americans, and French, the Chinese adapted an intermediate range ballistic missile called Long March 1 to carry their less-lethal space payloads. The last country to date to develop its own satellite and launching capability was the United Kingdom. The Black Arrow launcher, created for just this purpose, boosted the satellite *Prospero* into orbit on October 28, 1971. It was the only satellite launched with this British-made rocket. Since then, the British have relied on NASA launch vehicles for their various space projects.

All six countries entered this exclusive club to some extent for political reasons; for some practical and scientific motives were more important. For the Soviet Union, the United States, and the People's Republic, military reasons certainly figured highly. In the Soviet Union, there are two space programs, one military and one scientific. Military organizations apparently control the manufacture of all launch vehicles and supervise the launch facilities and operations. America's space program is more neatly compartmentalized. The National Aeronautics and Space Administration was created in 1958 as a civilian space organization, with the congressional mandate to promote the peaceful exploration and investigation of space. The Department of Defense, primarily through the Air Force, was left to conduct the country's military space program, the full details of which are not generally understood because of national security restrictions on the release of information. France, the People's Republic, and the United Kingdom all operate their spaceflight programs through the military, but civilian agencies develop much of the hardware and conduct most of the research. In Japan, of course, the entire program is in the hands of civilians.

Space Activities

Spaceflight, especially with orbital spacecraft, has opened entirely new vistas for the world's scientific community. Table 2 presents a record of space launchings successful in attaining Earth orbit or beyond. Although perhaps no radical changes in our theories about the creation and design of our solar system have resulted from our explorations of space, scientists do have a wealth of new data by which to understand planet Earth, its Moon and sister planets, and the medium of interplanetary space. Hundreds of investigations—astronomical, biological, geophysical—have been launched since the late 1950s. In addition to serving the scientists as information gatherers, satellites have been put to other uses. Surveying the planet from high altitudes, satellites serve as a tool for specialists who hope to improve the management of our natural resources and to increase the efficiency of agricultural practices. But it is sophisticated weather forecasting and communications that particularly attract new customers to the spacefold and keep them there.

Long-range weather predictions and high quality communications over long distances are two important, highly visible, practical contributions the space age has brought us all. The Soviet Union, Japan, and China in particular have important requirements for improving their communications and meteorological systems. Russia's and China's huge land masses make it difficult for them to develop adequate land-based communications systems and weather reporting networks at reasonable costs. Widely scattered communities can be connected through satellite communications links and weather patterns for large areas observed more efficiently from Earth orbit than from the ground. Both countries hope to bypass the complex ground-lines communications systems that serve the United States, Europe, and Western Russia by investing in satellite systems instead. For a crowded island population like Japan, reliable weather prediction is critical to agriculture, fishing, and personal safety. The Japanese have already developed an advanced communications

Table 2. World Record of Space Launchings Successful in Attaining Earth Orbit or Beyond*

Year	United States	USSR	France	Italy	Japan	People's Republic of China	Australia	United Kingdom	European Space Agency
1957		2							
1958	5	1							
1959	10	3							
1960	16	3							
1961	29	6							
1962	52	20							
1963	38	17							
1964	57	30							
1965	63	48	1						
1966	73	44	1						
1967	57	66	2	1			1		
1968	45	74							
1969	40	70							
1970	28	81	2	1	1	1			
1971	30	83	1	2	2	1		1	
1972	30	74		1	1				
1973	23	86							
1974	22	81		2	1				
1975	27	89	3	1	2	3			
1976	26	99				2			
1977	24	98			1				
1978	32	88			3	1			
1979	16	87			2				1
Total	743	1250	10	8	15	8	1	1	1

¹Includes foreign launchings of U.S. spacecraft.
*Note: This tabulation enumerates launchings rather than spacecraft. Some launches did successfully orbit multiple spacecraft.

satellite network that enhances their undisputed success in the fields of electronics and automation. In both the U.S. and Japan, business and industry have increased their use of facsimile and computer data transmissions, creating the so-called electronic office. Satellites play an essential role in this latest communications revolution.

The Soviets launched their first communications satellite, *Molniya 1-1*, in April 1965. Since that time through 1979, they have orbited 45 *Molniya-1*, 17 *Molniya-2*, and 12 *Molniya-3* class satellites, all of which had 12-hour orbits. In addition, they have sent three Gorizont, four Ekran, and three Raduga type communications satellites into 24-hour orbits to use for telephonic, telegraphic, television, and radio transmissions. In 1978, two amateur radio communications satellites called *Radio* were boosted into orbit. These two spacecraft were similar in purpose to the American ham radio satellite series known as *Oscar*. In 1978 and 1979, the Soviet Union also launched 54 military communications payloads as part of the Kosmos program; 48 of these were launched in groups of eight with six launch vehicles (Kosmos 976-983, Kosmos 1013-1020, Kosmos 1034-1041, Kosmos 1051-1058, Kosmos 1081-1088, and Kosmos 1130-1137). During the same time period, nine Kosmos navigation satellites were deposited in Earth orbits. The Meteor weather satellite program has included 27 Meteor-1 and 5 Meteor-2 class spacecraft.

By comparison with the Soviet Union and the United States, Japan is just beginning to build up its applications—or practical—satellite program, but it is moving ahead steadily. Japanese goals include the development of launch vehicles capable of placing satellites into geostationary orbit, the necessary tracking and control technology for such spacecraft, and the perfection of attitude control systems technology. NASA has launched two geostationary communications satellites and one geostationary meteorological satellite for the Japanese. Their first two attempts to orbit their Experimental Communications Satellite with the N-rocket in February 1979 and February 1980 resulted in failure. Concerned but undeterred, space agency managers and designers will continue with their program for a more advanced communications satellite system. A second Geostationary Meteorological satellite, GMS-2, is scheduled for launch by an N-rocket this year. In another applications program, the Japanese recently conducted an experiment in processing materials (an alloy, in this case) in space.

Space activities in the People's Republic of China are moving slowly from the initial stages of experimental launches and satellites to a more comprehensive program that will stress the practical applications of space technology, especially in communications, meteorology, and Earth resources management. In November and December 1978, Chinese and American space officials met in the U.S. (the American delegation led by NASA Administrator Robert A. Frosch and the Chinese team by President of the Chinese Academy of Space Technology Jen Hsin-min) to explore ways in which the two countries could cooperate in the field of space technology. A key topic in these discussions was the development of a civil communications satellite system for mainland China. Involved is the purchase by the Chinese of an American satellite communications system, including the associated ground receiving and distribution equipment. NASA would launch the satellites into geostationary orbit, and China would take over once the system was operational. A similar cooperative agreement was reached concerning the sale to China of a ground station capable of receiving Earth resources information from the NASA-National Oceanic and Atmospheric Administration Landsat remote-sensing satellites, including the Landsat-D scheduled for launch in the last quarter of 1982.

Since the first round of visits in 1978, the Chinese and Americans have had additional traveling exchanges involving government space agency officials and industry representatives. It is important to note that having successfully orbited domestically built satellites with their own launch vehicles, neither the Chinese nor the Japanese find it unacceptable to acquire foreign assistance with projects of immediate importance as they work to advance the state of their own technology—a very pragmatic attitude. China's most immediate goals are to develop a more powerful, efficient launch vehicle, advanced solid-state electronic components, and sophisticated communications and meteorological satellites.

China's new three-stage launch vehicle, called Long March-3, is expected to be flown this year, probably with an experimental communications satellite. The third stage of this vehicle will have a liquid-hydrogen and liquid-oxygen fuel system similar in concept to the American Centaur upper stage. These cryogenic fuels are difficult to handle, and the mastery of such technology by the Chinese will be a great leap forward. A 19-member delegation from the American Institute of Aeronautics and Astronautics visited China's aerospace facilities in November 1979 and made some candid assessments in their *China Space Report:* "We conclude that the Chinese are serious about their stated goal of an independent capability in communications satellites in the next decade, and are making good technological progress toward it. Their own frequently cited description of their technology as "primitive" is excessively modest. "Advanced, but simple," would be more apt. What they do lack, want, and expect to get from the U.S. is integrated know-how or "how to put it all together." They do not have experience or skills in systems engineering and program management. They do not seem to know much, for example, about designing to conflicting goals, such as performance, weight, power, cost, etc. They need information about reliability modeling and quality assurance techniques, and about scheduling and project control. To some extent the Chinese economic and social system has insulated designers from the concept of cost, at least for their own developments." Unfortunately, the Chinese have been forced by economics to postpone for several years the acquisition of the American-built satellites (two operational and one backup at about $150-250 million), but they will undoubtedly continue with their own research and development, even if at a lower level than before. Likewise, they have had to push back plans for their manned program until the 1990s (the first flights had originally been planned for the late 1980s).

If we tally up the total number of spacecraft launched from 1957 through 1979, we see that the USSR has a clear lead at 1,250. The U.S. follows at 743; then France 10, Japan 15, China 8, and the United Kingdom 1. Because of the Soviets' use of the catchall designation "Kosmos" (1147 of which had been launched through 1979) and the secrecy surrounding military satellites, we cannot classify all 2027 satellites by payload (scientific, meteorological, communications, etc.), but we can see certain trends (see tables). There has been an increase in communications and meteorological payloads over purely scientific investigations. Military payloads also—presumably many of these are communications and reconnaissance satellites—have been popular with the Americans and the Soviets. As public funds available for expensive space projects become scarcer in the years immediately ahead, it is probably safe to assume that ventures with some practical application that can be easily justified—like communications, weather forecasting, or military reconnaissance—will be funded more readily than scientific or experimental advanced systems payloads.

The Future

It can be dangerous for historians to venture into the field of projections; our crystal balls are as foggy as everyone else's. But the comments presented here are based upon projections made by Soviet, Chinese, and Japanese space experts. Clearly, there will be only five major space powers during the remainder of this century: the Soviet Union, the United States, the European Space Agency, Japan, and the People's Republic of China. And they will all apparently be concentrating their efforts on

Earth orbital operations for the foreseeable future, with occasional planetary probe missions for scientific investigation. All five powers look forward to their first manned or next-generation manned projects. The Soviets will continue with their *Soyuz-Salyut* missions, building toward a large Earth-orbiting space station. Americans hope to enter a new era of manned spaceflight next month with the launch of the first Shuttle orbiter. Shuttle flights will give European mission specialists assigned to ESA's Spacelab an opportunity to experience spaceflight, and the Japanese, among others, plan to send their payloads aloft via the new American space transportation system. Although the Chinese and Japanese cannot expect to conduct their first manned missions until late in this century, Chinese publications illustrate astronaut training in spacecraft cabin mockups, simulators, and centrifuges.

In the sphere of satellite projects, the Soviet Union will continue with its scientific, communications, meteorological, and military projects, with greater emphasis on Earth resources and oceanographic investigations. *Bhaskara*, launched on June 7, 1979, was a joint Soviet-Indian Earth resources satellite, and *Kosmos 1096*, launched on April 15, 1979, was believed to have been a partially successful ocean reconnaissance satellite (orbit decayed November 24, 1979). The Japanese are committed to launching increasingly advanced communications and meteorological spacecraft, but they also plan to become more deeply involved in Earth resources investigations and other practical missions, like material processing. For the mid-1980s, they have plans for biological payloads and limited lunar and planetary exploration with spacecraft of their own design and construction. Chinese plans call for the launch of their experimental communications satellites in 1981 and an experimental meteorological satellite the next year (the Chinese weather satellite has been described as roughly equivalent to the American Improved Tiros Operational Satellite—ITOS). This spacecraft will be placed in a 900-kilometer polar orbit. It is also likely that the Chinese will continue work with military reconnaissance satellites, and it has been suggested that their manned "Skylab" will have a military reconnaissance function; the same thing has been said for the Soviet *Salyut*. Manned observation craft could precede the availability of spacecraft equipped with remote-sensing devices by several years. A "box score" of space activity through December 31, 1979 is given in table 3.

Obviously, spaceflight is here to stay, and we will see the tempo of activity increase considerably in the coming decades. As Walter A. McDougall has noted, just as aircraft were the measure of a nation's technology between the two world wars, space technology has become the

Table 3. Space Box Score Through December 31, 1979

Country	Manned Activities		Unmanned Activities						
	Earth Orbiting	Lunar	Physics and Astronomy	Lunar and Planetary	Life Sciences	Meteorology	Communications and Navigation	Earth Resources	Military
Soviet Union	41	0	31[2,3]	52	6	29	74	1	Kosmos 1,147
United States (NASA and USAF)	22[1]	9	218[4]	39	8	47	114	4	374
France	0	0	15[5]	0	0	0	0	0	0
Japan	0	0	14	0	0	16	3[6]	0	0
China	0	0	7	0	0	0	0	0	1
United Kingdom	0	0	1	0	0	0	0	0	0

Notes: All categorizations are approximate. For example, the Soviet Kosmos series includes many scientific spacecraft, but the Soviets generally have not given details on their projects.

[1] Includes 2 Mercury-Redstone suborbital missions.

[2] Includes engineering test spacecraft, but does not include scientific satellites flown in Kosmos series.

[3] Includes joint missions flown with France, India, and the Warsaw Pact Nations.

[4] Includes joint missions flown with Australia, Canada, ESRO, Federal Republic of Germany (FRG), France, Italy, Japan, NATO, Netherlands, and Spain.

[5] Includes joint missions with USSR, USA, and the FRG.

[6] Includes 1 meteorological and 2 communications satellites launched for the Japanese by NASA.

post-1945 symbol of technological prowess. Although the spaceflight enterprise began as an extension of Cold War competition and scientific inquisitiveness and grew mightily because of the power and prestige it brought its backers, it has been sustained for its practical values, for its everyday utility. To be certain nations will continue to measure one another by what they have or have not accomplished in certain technological arenas, and space will be one of them. But individual nations will examine their own activities in terms of the practical benefits their space programs are bringing their own people and socioeconomic system. Space may still be the "high frontier"—with all the hope and adventure that that term implies—but it is the dividends delivered back to Earth that will keep the adventure going.

Source Notes and Recommended Reading

Soviet Space Program:

Riabchikov, Evgeny. *Russians in Space*. Translated by Guy V. Daniels. Garden City, NY, 1971 (an official view of the Soviet space program prepared under the direction of the Novosti Press).
Smolders, Peter. *Soviets in Space: The Story of the Salyut and the Soviet Approach to Present and Future Space Travel*. Translated by Marian Powell. Guildford and London, 1973.

Japanese Space Program:

Keidanren [Federation of Economic Organizations]. *Space in Japan, 1978–79 (Tokyo, 1979)*.
Kuroda, Yasuhiro. *"Overview of the Japanese Space Activities," paper presented at the 1979 Australian Astronautics Convention (Perth, August 20#25, 1979)*.

Chinese Space Program:

Pritchard, Wilbur L., and James J. Harford, eds. *China Space Report: An Eyewitness Account of China's Space Activities by a Delegation from the American Institute of Aeronautics and Astronautics, November 1979 (New York, 1980)*.

General References:

TRW Defense and Space Systems Group. *TRW Space Log*. Published by TRW's public relations staff. (Redondo Beach, California, 1960).

COMMENTARY
Richard S. Kirkendall

This session looks at its subject—"Domestic and International Ramifications of Space Activity" or "The Politics of Space"—from several perspectives. They include two disciplines, both political science and history; and seven nations, the United States, France, Great Britain, West Germany, the USSR, the People's Republic of China, and Japan. And the people who do the looking have had rich experience in research and other relevant activities.

For me, these are very useful papers presented at a crucial stage. As a new member of the advisory committee of the NASA history program and a historian who has specialized in cultural and political approaches to recent history rather than the history of science, technology, or the space programs, I need the education that these experts in the history of space programs supply. I hope the papers are equally useful to the members of this audience. In hope of enhancing their usefulness, I will summarize them, stressing the interrelations among them.

Professor Logsdon is concerned chiefly with the principles that have governed the American space program and the institutional expressions of those principles. He defines six that came to dominate in the crucial years from 1957 to 1962, the years of President Eisenhower and Kennedy and of *Sputnik* and the decision to go to our Moon. The principles are:

1. Activities in space can be justified by political as well as other objectives—scientific, military, intelligence, economic.
2. The United States should be preeminent in all areas of space activity.
3. Civilian and military space activities should be separated.
4. NASA should be limited to R&D.
5. The government should encourage private-sector involvement in the use of space technology but should itself sponsor research in areas of potential commercial application.
6. National objectives rather than international cooperation should be in first place.

According to Logsdon's very skillful analysis, some but not all of these principles are still in control. No longer so heavily influenced by political considerations in this area, the United States no longer insists upon preeminence. Civilian and military activities continue to be separated from one another; NASA remains confined to research and development; the government still sponsors research that could lead to profitable commercial operations. As to the relative importance of nationalism versus internationalism in space activities, the present is less clear than the past, while the future is uncertain.

Professor McDougall deals with a very different program. Compared with the American program, the Western European is small and has produced small results. He sees it as an illustration of the history of Western Europe since World War II. Once the leading area of the world, it has, since the 1940s, been dwarfed by the Soviet Union and the United States. In space, it has responded to the accomplishments of both of the superpowers in French-led efforts to reassert itself, but this once-powerful and still proud part of the world has accomplished very little.

Provided as we are with the opportunity to see both the American and Western European programs, we can see some resemblances. The latter has been influenced by political ambitions, represented especially by the French. It has been influenced also by a sense of limits, with the British role most important here. And the West Germans especially have demonstrated scientific and technological capabilities similar to those in the American program. Also, in Western Europe as well as in the United States, we see that coordination in space efforts can be difficult to achieve, the relations between the public and the private sectors are not easily defined, and there is tension between international cooperation and national self-assertion. In both programs, businessmen testify to their eagerness to benefit from government activities and political men try to use businessmen; government organizations develop and use business organizations.

The Western European space story also provides a dramatic illustration of the significance of DeGaulle. It even encourages one to suggest that Ronald Reagan may be the American DeGaulle, however offensive some may find such an analogy. And the story suggests that it is not easy to reverse decline.

The first paper deals with the second member of the "space club" or the "Space Six," as Dr. Ezell labels the nations that have launched satellites. The second paper focuses on the third member (France) and the sixth (the United Kingdom). Ezell examines the first (the USSR), the fourth (Japan), and the fifth (the People's Republic of China). In the process, he supplies some useful statistics on the sizes of the different programs. They illustrate how much larger the Russian and American programs are than the others.

Ezell makes other contributions. He adds to our understanding of the complexity of the participation in the space programs, and the roles of civilian and military organizations and of the public and private sectors.

He shows that the pattern of participation varies from country to country, with the military role, apparently, largest in Russia, smallest in Japan. He underscores the importance of political considerations but makes it very clear that they have not been the only influences, that they are weaker now than they were at first, and that practical benefits are now very influential. And of the three authors, Ezell seems most optimistic about the possibility of international cooperation in space programs.

Ezell makes the largest effort at prediction. He concludes that "space flight is here to stay, and the tempo of activity will increase considerably from this point in time to the end of the century." Logsdon, with his sense of being an early student of a chain of events of great significance must agree with this prediction. "Working on space history," he writes, "is one way for those of us without high technical competence to get close to what is (to me at least) the great adventure of my lifetime." "Future historians," he adds at the end, "are almost certain to view mankind's first tentative expeditions away from its home planet as major historical events. From this perspective, it is a privilege to be in at the beginning." Forced to deal with Western Europe, McDougall does not have the same opportunity to express a similar sense of personal significance. But his broader efforts, his use of the term "early space age," suggest that he has such a sense.

The prospects before the world, one might conclude from these papers, resemble those facing Western Europe as the 15th gave way to the 16th century. Now, however, Western Europe is not in the strong position to exploit the opportunities that are opening up that it was five centuries ago. These papers encourage us to think in such terms.

The papers are important for the methodology involved as well as for the information presented. And for this we are the beneficiaries of the designers of the conference as well as the presenters of the papers. The designers put together a session that enables us to see the benefits of the comparative approach to history. Those benefits seem to me, in the case of the history of space programs at least, to be very large.

The session supplies, of course, a preliminary and not all together explicit or conscious exercise in comparative history. The session, organized as it is, helps us see similarities and differences in the various programs, but the papers do not consistently call our attention to those similarities and differences or attempt to explain them. But we should be grateful that the session goes as far as it does for, as George M. Frederickson has observed in *The Past Before Us* (p. 472): "When all is said and done. . ., the dominant impression that is bound to arise from any survey of recent comparative work by American historians is not how much has been done but rather how little." The American historical profession, he suggests, is not organized in ways that encourage comparative work. Perhaps the discussion from the floor can push us farther toward a comparative history of the space programs.

THE RATIONALE FOR SPACE EXPLORATION

THE IDEA OF SPACE EXPLORATION

Bruce Mazlish

In the 1950s, man first ventured into outer space. At the end of the 1960s, he was on the Moon, having traveled over 200,000 miles and at speeds upward of 18,000 miles per hour. The modern Daedalus had taken his first step into reality. An age-old dream had been realized. A proud Wernher von Braun compared it to that moment in evolution "when aquatic life came crawling up on the land."

Now we seem to be crawling back. The Moon landing, for all the impact it had during that sultry July night in 1969, has scattered into small effects upon us. Our expectations fulfilled, we now seem to have lost interest. I am puzzled by the disparity between the greatness of the deed and the meanness of the result. How to explain it?

To explore further the gap between the deed and its estimation, we can proceed along two major paths: to compare space with past episodes of exploration and development; and to examine the contemporary context in and of itself. Both, even briefly examined, are revealing.

In comparing space with past episodes that bear a resemblance to it, we are engaging in historical analogy. Historical analogy gives flesh to a perception of vague resemblance. It is not a rigorous form of reasoning, but it is one of the more attractive. It is, too, a fashioner of myths—durable ones that survive, like a locust's brittle armor, even after life itself has departed. Analogy, finally, has but one eye, and it sees only similarities.

The analogy that immediately springs to mind is the Age of Discovery. One is struck by the similarities: a desire for national prestige; a hope of gain, both economic and military; an impulse to adventure; sheer curiosity. There also was a religious factor in the 15th century. Even that finds a 20th-century expression in our notion of scientific "mission."

In the end, however, I do not believe that the analogy of the space program, emphasizing its exploratory aspect, with the Age of Discovery is as useful as some others (e.g., with the railroad, as I shall attempt to show). We have inaugurated an age of discovery, but it is not *the* Age of Discovery, and it lacks the props and resonance we were conditioned to expect.

The major difference, I believe, is that in space there are no flora and fauna. There are no people on the Moon to be conquered or converted. There are no new animals to grace the parks of a Spanish king, no exotic

137

plants to nurture in the royal gardens at Kew. Columbus returned with naked savages. Lewis and Clark identified 24 Indian tribes, 178 plants, and 122 animals, all of them previously unknown. Even the voyagers of the Beagle sailed into port with exotic, if ugly, Fuegians that titillated the English public.

Space, in comparison, is "empty," and our chief harvest thus far has been in the form of rocks. The Moon is unpopulated; its "man," visible from 200,000 miles away, vanishes on close approach. The only earthly comparison is the arctic and antarctic, although they are, in fact, more richly endowed, and neither of these, for comparable reasons, has ever aroused much enthusiasm. Vast, cold worlds, they lie largely untapped and unsettled.

How can one become enthusiastic about such "inhuman" areas? Exploration of such "terrains" cannot give rise to a sense of "climates of opinion," which shake the traditional order. It does not leave us with the 19th century's feeling of being "Between Two Worlds," either in time or geography. Where early explorations were preceded by myths about gargoyles blowing off shore, or apes raping women (as Voltaire fondly imagined), or even abominable snowmen, the main equivalent titillation of the space effort was a scientific surmise about the possibility of some kind of extraterrestrial life. In this, we were soon to be disappointed.

In such an empty world, devoid of any presence other than one's own in a clumsy, bulky spacesuit, myths and imagination crumbled into computer bits. The symbolic nature of space dissolved. Physical and biological scientists might well be absorbed, but what was there to interest their social science and humanistic colleagues? Or the general public, for whom the latter served as interpreters?

If space and the Moon offered so little of "human interest," what of the explorers themselves? They, too, failed to capture our imaginations. They were fighter and test pilots turned astronauts, but not adventurers. They were not heroes, in spite of NASA's media hype (and though the age was antiheroic, it was ambivalently so). Instead, the astronauts were a team, replaceable men, with not a Columbus or even an Amerigo Vespucci among them. The Moon landing craft might be called the "Eagle," but no Lindbergh, in lone splendor, sat at its controls. The argument over manned and unmanned spacecraft was without "human" consequence, for the astronauts became replaceable and duplicable instruments just as much as the unmanned vehicles.

Norman Mailer, in one of the few attempts to respond imaginatively to the space effort—one thinks earlier of Camoen's *The Lusiads*, or Shakespeare's *The Tempest*—brilliantly attempts in *Of a Fire on the Moon* to kindle sparks of imagination to set aglow our hearts and minds.

He speaks of dreams that border on either madness or ecstasy, of Hemingwayesque courage, and dread of death. All to almost no avail. NASA, in its very concern that an *Apollo 11*-connected death would result in the end of support for space investigation, unknowingly aborted the public's interest. As Mailer puts it, "The irony was that the world, first sacrifices in outer space paid, would have begun to watch future flights with pain and concern." Death fears and dreams gave way to a TV picture, whose dramatic appeal was almost nil. Tranquillity Base took on, unintendedly, a soporific quality that spread out over the entire space program. So much for the Age of Discovery analogy.

The other major analogy useful to make is with what elsewhere I have called "social inventions." [1] I define it as an invention that is technological (e.g., missiles, launching pads), economic (e.g., involving large-scale employment of manpower, widespread use of materials), political (e.g., involving new forms of legislation, and new dispositions of political forces), sociological (e.g., affecting kinship groups, communities, classes), and intellectual (e.g., changing man's views of space and time). Such an invention has a profound effect on us; it is literally "revolutionary." The lowly cotton industry in the early 19th century and the railroad in the mid-19th century, in Britain, were of this nature. Thus, the innovations in cotton manufacturing had enormous secondary and tertiary effects, helping to spark the Industrial Revolution, or what W.W. Rostow has called "sustained takeoff": cotton manufacturing brings into being the factory, and its operatives (or proletariat, a new class); groups the workers in an increasingly populated urban setting; stimulates the growing of cotton and the cotton trade (not to mention the slave trade); and strongly affects the coal and iron industries by its demand. A Manchester, as well as a Manchester School of Free Trade, symbolizes its impact. There is no comparable "Manchester"—Cape Canaveral will not do—in space development.

The railroad is of a similar magnitude to cotton manufacturing, but more analogous to the space program in its use of engines for transportation, though without the element of exploration. The railroad, like the space program, for a while also annually consumed about 2 ½ percent of the GNP as its investment requirement. But think of the railroad's impact on communities, on social structure, on related technologies, on the economy as a whole in comparison to the space program, i.e., its return to society!

And now remember the optimistic predictions. In 1963, Robert Jastrow and Homer E. Newell predicted that the space program would mean "the benefits of basic research, economically valuable applications of satellites, contributions to industrial technology, a general stimulus to education and to the younger generation, and the strengthening of our international position by our acceptance of leadership in a historic enterprise." Erik Bergaust exalted: "Fifty years from now? Who knows, perhaps we will terminate the use of the title doctor—because everyone will have at least a Ph.D. degree. That might well become a typical result of our current Space Age brainpower drive." Toby Freedman, Director, Life Sciences, North American Aviation, Inc., announced that in his own field of "medical miracles," contributions exist "that to my mind have already paid back the cost [of the whole program]."

Critics of the program, on the other hand, point to its huge costs—40 billion dollars plus for Saturn, 12 billion dollars alone for the construction of the Space Shuttle, and another 15 billion dollars projected to operate it—and ask whether the touted side effects of the space program could not have been achieved directly and more effectively by the expenditure of lesser sums of money. Most of us want less "spaced out" reasons for spending the enormous amounts involved to loft such massive payloads as *Saturn V/Apollo 11*, with such seemingly minuscule payoffs, whether in material benefits or psychological rewards.

If anything, the overblown claims of space enthusiasts have come back to haunt them and to add to public disillusionment. Wayne Biddle is typical when he concludes his article on the Space Shuttle[2] by detailing its problems, as much political as technological, and saying that "the real driving force is clearly not the solid promise of cheap, routine access to space." Space exploration, in short, has not revolutionized our lives, or any part of them, though it is clearly powered by mundane as well as purely scientific motives.

The justification in terms of national prestige today fares no better. We see an American space program, whose liftoff took place as a result of the Cold War. The impetus in 1957 was clearly rivalry with the Soviety Union; that was justification enough for huge expenditures. Earlier explorations, e.g., in the 16th century, did result in military conflict. Macabre as is the thought, even a small-scale conflict in space would rivet public attention on the program. Science fiction is filled with such wars—and hence "human" interest: we think of the movies, "Star Wars," and the TV shows, "Star Trek" and "Battlestar Galactica." (Incidentally, "Star Wars" also appeals because of its peopling outer space with strange other humans and with imaginary animal-beings.) Our more fortunate and peaceful present lacks such daring, and pays the price in public boredom with space. In addition, with the change in public opinion after the Vietnam war, plus our Pyrrhic victory in the space race—how has this really advanced us against the Russians? The military and national prestige motive has lost much of its force.

What is left? The "high" has been taken out of the adventure—a humanless space and a heroless program have seen to that. There are no heathen to missionize, no or little further military and national prestige to be gained immediately, and either paltry or very long-range economic gains to be reaped.

What is more, space science has been caught up in the same revulsion that has manifested itself so strongly against general science in our contemporary culture, a revulsion whose symbolic expression has become the nuclear protest. True, the revulsion is flamed by a small, activist group, while the general public remains silently supportive of science, as polls show. But the activists have made physics and its kin appear as a Pandora's box more than a cornucopia. The "Idea of progress" has lost its automatic conviction.

The forces justifying space exploration, therefore, have become discretionary. As a discretionary matter, and not a matter of unquestioned national purpose, the space program is now weighed against other discretionary expenditures—cancer research urban renewal—often found wanting and wasteful by comparison. Until space colonization or stepped-up military conflict in space come along to rekindle public interest, the space program's chief ally seems to be leftover momentum: the fact that certain programs, planned long ago, happen to be under way.

Yet, to my mind, there are two arguments that suffice to justify a leap into space, both of them as unprovable as they are irrefutable. The first is that the flight into space changes our whole view of ourselves and the Earth. The fact of sheer flight itself, while enormously significant, is not of the same order of importance. One could, of course, say, "Well, the spacecraft is simply an extension of the airplane. Man has flown already, and that's the big breakthrough." In part, this argument is correct: by leaving the Earth in sustained flight, even if only 20 feet off the ground, man changes his nature, extends it to the aves class. Within a few decades of Kitty Hawk, Hubert Wilkins, later Sir Hubert, flew over the barren wastes of the Arctic and Antarctic, followed by Richard Byrd over the North and South Poles. Armstrong and Aldrin flying past equally barren wastes on the Moon, even setting foot on it, in this sense do nothing new.

The newness, the greatness, resides in the fact, not of flight, both of man's thrusting himself out into space past his terrestial abode and the atmosphere that has nourished and protected him. As Hannah Arendt noted, man now occupies a position from which he can observe his own abode as an "outsider," both physically and philosophically, poised to explore further the rest of his solar system—and beyond. It is not the mechanical flight, awesome as that is, but the spacial reorientation, mental as well physical, that marks the new evolutionary step.

Put very simply, the Earth is now perceived as itself a spaceship. Suddenly, all Earth is turned into a larger form of the very vehicles it sends into space—a macrocosmic form of the microscopic projectile that is powered into a fixed orbit. The Earth is now conceived of as a "ship" navigating the "ocean" of space, carrying its human crew and their life-sustaining equipment.[3] Now, too, there is the sense that the ship, Earth, can go down, i.e., be shipwrecked. Only in this case, it will have been the human crew, not the oceans of space, that innundate or befoul the ship, and thus wreck it.

The Earth as spaceship, therefore, is a newly imagined way of conceiving our terrestial abode. A comparison with previous attitudes toward "Mother Earth" shows how the conception of a "spacecraft" frees us—in a terrifying way—from the old reassurances embodied in the notion of *terra firme*.[4] The whole Earth has become Daedalus—with no fixed landing place, psychologically, to which to return from its flight.

The second argument justifying the space program is that it is man's destiny continually to test himself against the unknown, to know himself by his exertions. And to my defense I call upon an earlier traveler in unknown spaces, Ulysses, encountered by Dante in the Inferno:

> "O brothers," I said, "you who
> through a thousand perils have come to the West,
> to the brief vigil of our senses
>
> which is left, do not deny
> experience of the unpeopled world
> to be discovered by following the sun.
>
> Consider what origin you had;
> you were not created to live like brutes,
> but to seek virtue and knowledge."

Source Notes

1. *The Railroad and the Space Program: An Exploration in Historical Analogy*, ed. by Bruce Mazlish (Cambridge, MA: MIT Press, 1965).
2. *New York Times Magazine* (June 22, 1980) p. 40.
3. The romantic depiction of a "spaceship" returning from the Moon by the French illustrator Gustave Dore (1833–83) [in *The Wilson Quarterly*, Autumn 1980], with its craft an actual sailing ship in the sky, halfway between the Moon covered by scudding clouds and the heaving waves of the terrestrial ocean, graphically links the images of ship, sea, and space—and rightly reminds us of the pull of the Moon upon the tides, thus connecting the two "worlds."
4. One form our anxiety has taken is in the "sighting" of UFOs. They can be explained, psychologically, as a projection of our own intrustion into space—we project our intentions and actions unto others. (For a fuller analysis, see C.G. Jung's article, "Flying Saucers: A Modern Myth of Things Seen in the Skies," in *Civilization in Transition*, vol. 10 in the Bollinger Foundation series of the collected works of Jung.) Of course, earlier centuries, too, have always assumed interventions from heaven, but these were in the form of gods, plagues, etc. The UFOs, naturally, mirror our current beliefs.

A HUMANIST'S VIEW

James A. Michener

The organizers of this conference have done me great disservice in designating me a humanist, because there is an army of charismatic preachers throughout the nation who say they are going to drive all humanists from public life. We are held to be subversive to family solidarity, destructive of honest religious principles, and committed to crypto-communism.

I do not recognize this description of myself. The humanists I think of when the word is said are Sir Thomas More, William Shakespeare, Johann Goethe, Plato, and Benjamin Silliman of this university. I have always strived to follow decently in their footsteps, and I am proud to call myself a would-be humanist. Let me review how one humanist, over a course of nearly 75 years, has responded to the challenge of space.

As a young boy I had the good luck to acquire a copy of *Norton's Star Atlas* published by Gall and Inglis of Edinburgh. With it, and especially its small-print text, I first explored the heavens, and through the decades I have always kept a copy with me.

In World War II, I served in Guadalcanal and New Zealand, where I was able to study with special care the southern celestial hemisphere.

As a consequence of such investigation I became interested in cosmogony and starting in 1948 began to read all that appeared on this subject as it was published. I became a devotee of Fred Hoyle's theories and constructed a rational scenario for a closed universe that was constantly replenishing itself. The more I studied, prior to 1960, the more satisfied I became with my theory.

Why was I, a nonastronomer, interested in cosmogony? It seemed to me then, as it does now, that a certain percentage of the human race is obligated to speculate about ultimate cause, for from such speculation ensues great understandings. In ancient Assyria I would certainly have studied the stars. At Stonehenge I would have helped align the stones at the solstices. In medieval Poland I would have been agitated by the theories of Copernicus—and as a traditionalist would have opposed them; but Newton would have blasted my mind loose, especially when I compared his revelations with what Kepler had been saying. I would have waited avidly for each new report of the telescope astronomers.

Now we come to what I actually did. Wherever I have lived I have spent a year marking the solstices and the elevations of the Sun, which is what any prudent man ought to do. I studied Percival Lowell as carefully as a layman could and concluded that he was talking nonsense. On the eve of *Mariner 9's* marvelous photographic revelations I stated publicly my conviction that Mars would produce no ascertainable life and repeated my opinion at the colloquium held in 1976 on the eve of the actual landing on Mars. I tried my best to visualize the planetary system.

During the first years of public discussion of Einstein's theory I was totally confused but with the aid of certain elegant expositions I worked my way to a layman's understanding, and it has been a joy to follow subsequent ramifications, which acquire special significances today.

In 1965, my comfortable assumptions were boldly shaken by the work of Penzias and Wilson. They informed the world that their massive antenna was picking up radiation which could not be accounted for by known centers of emission. Other investigators suggested that this must be cosmic blackbody radiation, at the predicted 3 K temperature and on the 3.2 cm wavelength. From this it was an easy leap to decide that this must be the residue of the Big Bang which astronomers had contemplated and predicted.

All my early conjectures were blown apart, and I was forced to think of an unlimited universe. Quasars with their tremendous distances and speeds required new understandings. Pulsars provided equivalent enigmas in the radio field and provided opportunities for radical new interpretations. Black holes gave me no trouble, for I had long speculated about the ultimate consequences implied by aspects of the Einstein theory. And the concept of a singularity, with all it implies, was not difficult to accept.

It was at this point that I became seriously interested in the work being done by NASA. I followed with care the mapping of the Moon, the sending of spaceships to the outer planets, the sending aloft of telescopes which could photograph astronomical bodies freed from atmospheric distortions. The stupendous additions to our visual understanding of the universe were of great significance, because I agree with the Chinese that one picture is often worth a thousand words.

On my desk these days I keep a copy of the amazing photograph of Quasar 3C-273 with its ejecta of staggering dimension. The other day I made some rough calculations and deduced that if this great out-thrusting is cylindrical, as it appears to be, it would provide room to contain 60 million billion of our earths (5.96×10^{16}).

And as I contemplate it, and the dazzling discoveries of which it is a minor part, I find that all interlock in their significance, and I become aware that mankind is in the midst of one of its noblest periods of intellectual expansion. I liken it to the Copernican Age, to the Age of Newton, and to the explosive consequences of Charles Darwin's theory of evolution.

And inescapably I have to ask: If this age is so tremendous in the implications of its discoveries, why is the general population so unaware of, or so indifferent to if aware of, the stupendous situation in which we find ourselves? We live amid a fantastic explosion of meaning and we remain almost indifferent to it. But was it any different in the past?

The Copernican Age. I have been doing much work in Poland and I calculate that not more than 1 percent of the persons living at that time could have heard of the Copernican discoveries. But I also find that those who were going to modify life in the centuries that followed *did* make themselves aware.

Newton's Codifications. These could not have been known by the vast majority of persons living with Newton, or have been understood. But the life of everyone was to be modified by the scientific consequences which were inspired by Newton's revelations. Again, those who needed to know, knew.

Darwin's Theory. This differed from the other two because it did produce an immediate fallout, since it touched religion. It occasioned heated public debate which continues. When I was in Alabama recently I had an opportunity to hear several of the new electronic ministers. They were a brilliant lot, remarkably able and persuasive. I found myself agreeing with some of their major points, as many sensible listeners would, but when they began to attack me as a humanist I shivered, for they said specifically that they intended driving people like me out of public life. Their attack was focused heavily on Darwinism, and by extension on geology, anthropology, paleobotany, and modern explorations of astronomy. It is entirely possible that the day might come when, if you want your daughter or son to explore the ultimate meanings of space, she or he might have to emigrate to Germany or Japan.

Tonight I stand confused. On the one hand, it seems as if our nation has turned its back on space exploration. One major program after another is being scuttled and we will probably not even send a messenger out to greet Halley's Comet, which comes our way once in 76 years.

- Congress is hesitant to commit money to the grand adventure.
- NASA has no clear mandate for the years ahead.
- The public is apathetic, even about the Jupiter and Saturn flybys.
- Russians are constantly perfecting their skills and understandings.
- As to the great discoveries, we live in an age of light but insist upon hiding in a cave.
- There is a constant and growing rebellion against science, witness: growth of astrology; rejection by young people; attacks by the clergy; attacks on Three Mile Island; response to the DC-10 problem; wild agitation over Skylab.

On the other hand, I see the public increasingly fascinated by space: the space museum in Washington; *"Star Wars"* and *"Close Encounters"*; acceptance and growth of science fiction; enthusiastic support for the work of Carl Sagan; explosive distribution of handheld calculators and home computers.

I can be excused if I am confused. In 1938, President Roosevelt assembled a seminar of the brightest American scientists available and asked them what radical developments in science the American government ought to anticipate. The scientists handed him a thoughtful report in which they failed to predict six startling developments about to explode on the scene: radar, penicillin, computers, jet aviation, rockets, and atomic explosions. How can our society anticipate and prepare for the explosive discoveries that loom ahead if we block orderly discussion, exploration, and experimentation?

What should the posture of NASA be at this critical juncture? I believe we must commit ourselves to the logical next steps in the exploration of the universe. Our strategy must be to prepare ourselves for physical exploration of the solar system and for the unmanned exploration of the remotest regions of outer space. But what should our practical tactics be?

- NASA should adjust easily and intelligently to such temporary missions as the Congress and the intellectual community can agree upon, on the defensible grounds that even a small step in the right direction is a worthy step.
- NASA should strive to sustain and enhance the nation's vision.
- NASA should maintain pressure for funding to support essential missions.

- NASA must, above all else, preserve the cadre of informed experts prepared to take the next steps. I deem this to be a national priority, and of the most austere necessity.

In these days of anxiety over lost opportunities and possible next steps, I often think that the United States is in the position in which Portugal and Spain found themselves in the 16th century. They had made the stunning explorations, but then they withdrew from the competition and stood aside as France and England took the next steps, including the explorations upon which nations and empires were built. The losses that follow upon such surrenders are inevitable and irreversible.

I have said that on my desk I keep a photograph of 3C-273 to remind me of the immensity of our universe. I keep another photograph in my work file, where I study it almost daily because I have long felt that it was the most beautiful object in nature. It is that stunning, clear, aloof portrait of N.G.C. 4565, an edge-on galaxy which must look much as ours would from a comparable point in space. It lies in Coma Berenices and is invisible to the naked eye or even to a small telescope. But there it rides, immensely far away, immensely beautiful, reminding us that we who ponder the problems of space and the universe are also involved in the meaning of beauty.

COMMENTARY

Carroll W. Pursell, Jr.

I have been asked to comment on papers exploring the rationale of the space program and I will take the word rationale to mean the underlying reason or the rational foundation. At the same time, I think it important to watch for what Bill Holley has called the *real* reasons, as distinct from the *good* reasons.

Mazlish identifies the common rationales for the space program: the age-old dream of space travel, the hope of revolutionizing our lives, the dream of economic payoff, and the quest for national prestige. To these he adds his own reasons: the concept of spaceship Earth has changed for the better our own view of ourselves and our world; and the destiny of humankind is to push against the frontiers of the unknown. Mr. Michener, in a graceful and moving personal testimony, invokes much the same sort of spiritual and intellectual imperative.

Both papers share common points. First, they favor science as such and do so largely because it is good, true, and beautiful. Such a view is not necessarily wrong but it provides no criteria for independent analysis. As the then Director of the Bureau of the Budget quipped about Vannevar Bush's famous 1945 report, it might as well have been titled *Science the Endless Expenditure.* The real problem is not whether or not science is good, but rather how much we can afford to buy, and for what purposes.

Another common assumption is that anti-science is a bad, and perhaps growing, phenomenon in our country today. Mr. Michener worried about the mood of the nation. I do not share this perception, however; recent polls show that the American people still hold science in high esteem. It is possible that technology has been somewhat demystified in recent years, but that is another matter and, in my own opinion, a good thing. If John Higham was correct in his idea that technology has been the most recent and powerful unifying agent in American culture, its fate is certainly a serious matter. However, neither of the speakers really raises that issue.[1]

I think that history tends to undermine the basic rationales for the space program, but history also should give us some reassurance about current changes in the fate of the entire enterprise. The analogy with the Age of Exploration is too simply put. It was not a period of rivalry only between Spain and England but between these two and Portugal, France, Holland, and others as well. The fact that the fortunes of each waxed and waned should be a source of reassurance, not of alarm. In the long run which of these has triumphed? None are today great superpowers.

Nor should one be unduly alarmed at the sudden threat to funding for space sciences. Science, like all other fields, has always been subject to "fads." Since the late 19th century, geology, chemistry, physics, and biology, in something like that order, have been the "hot" fields of science. None go away, all advance, and certainly science as such is not tied to any particular ranking among them. Like the example of the Age of Exploration, the lesson is not one of doom, but a caution against assuming that Western civilization rests solely on current (perhaps already eclipsed) institutions and enthusiasms.

I think that the basic ideological thread running through both papers, and perhaps through the entire space constituency, is very close to the motto of the 1933 Chicago Century of Progress world's fair: "Science Finds—Industry Applies—Man Conforms." [2] Dr. Sylvia Fries, chair of the NASA Historical Advisory Committee, has discovered in a recent study of congressional testimony on science and technology policy a pervasive and consistent belief in the notion that "technology is the instrument by which Man transforms science into history." This is an unexamined and debatable proposition, and any fear for the fate of civilization based upon it is an act of faith, not a commitment to rational progress.

The concern of our two speakers for preserving and supporting the good that underlies the space program is to be applauded and shared by all of us. What that good is, and how closely it must be tied to the space program itself, is a question we have hardly begun to ask.

References

[1] John Higham, "Hanging Together: Divergent Unities in American History," *The Journal of American History,* vol. 61 (June 1974), pp. 5–28.
[2] Lowell Tozer, "A Century of Progress, 1833–1933: Technology's Triumph Over Man," *American Quarterly,* vol. 4 (Spring 1952), pp. 78–81.

PARTICIPANTS' BIOGRAPHIES

PARTICIPANTS' BIOGRAPHIES

Edward Clinton Ezell is Curator of Military History at the National Museum of American History. After taking a Ph.D. in the history of science and technology from Case Institute of Technology in 1969, he taught at North Carolina State University and Sangamon State University. Under contract to NASA, he wrote with Linda Neuman Ezell: *The Partnership: A History of the Apollo-Soyuz Test Project* (1978) and *On Mars: Exploration of the Red Planet, 1958–1978* (1984). At the time of this conference, Dr. Ezell was the historian of Johnson Space Center. He has also published extensively on military technology, including *Small Arms of the World* (11th ed., 1977).

Sylvia Doughty Fries is an historian who received the doctorate from the Johns Hopkins University in 1969. The author of a number of studies in colonial and 19th century American ideas and attitudes, she is currently exploring the ideological makeup of our contemporary scientific and engineering professions. She has served as a historical consultant for the National Science Foundation, the National Endowment for the Humanities, and NASA. A former chair of the NASA History Advisory Committee, she is now Director of the NASA History Office.

Richard P. Hallion is historian of the Air Force Flight Test Center, Edwards Air Force Base, California. He received his Ph.D. in the history of American technology from the University of Maryland in 1975, and served as curator of Science and Technology at the National Air and Space Museum from 1974 until 1980. Among Dr. Hallion's many publications in aerospace history are *Supersonic Flight: Breaking the Sound Barrier and Beyond* (1972), *Legacy of Flight: The Guggenheim Contribution to American Aviation* (1977), *Test Pilots* (1981), and *On the Frontier: Flight Research at Dryden, 1946-1981.* He has taught in both the Department of History and the Department of Aerospace Engineering at the University of Maryland and has been a Visiting Fellow at Yale University.

I.B. Holley, Jr., is Professor of History at Duke University. After service with the Army Air Forces in World War II, Professor Holley took his Ph.D. in history in 1947 at Yale University. He teaches American social and intellectual history, but his major research and writing have been in military history and the history of technology. Best known for his *Ideas and Weapons* (1953, 1971), Professor Holley has also authored *Buying Aircraft: Materiel Procurement for the Army*

Air Forces (1964), and most recently *General John M. Palmer, Citizen Soldiers, and the Army of a Democracy* (1982).

Richard S. Kirkendall is the Henry A. Wallace Professor of History at Iowa State University. When he participated in the conference (and for seven and one-half years before that date), he was both Professor of History at Indiana University, Bloomington, and Executive Secretary of the Organization of American Historians. He has also taught at Wesleyan University (1955-1958), and the University of Missouri, Columbia (1958-1973), chairing the Department of History in the latter from 1968 to 1971. His publications include *A Global Power: The United States Since the Age of Roosevelt* (1979).

Arnold Levine has taught at Brooklyn College, served as project manager for a Washington, D.C., research organization, and is currently the marketing researcher for a large architectural and engineering firm in Fairfax, Virginia. Dr. Levine received his Ph.D. in European history from the University of Wisconsin in 1972. He has written several articles on the relations between education and scientific research and has delivered papers before the American Historical Association and the Sixth International Congress on Economic History. His book on *Managing NASA in the Apollo Era,* of which this paper is a summary, was published by NASA in 1982.

John M. Logsdon is Director of the Graduate Program in Science, Technology, and Public Policy of George Washington University, where he is also Professor of Political Science and Public Affairs. Holder of a Ph.D. in political science from New York University, Dr. Logsdon has pursued research in space policy, the structure and process of government decisionmaking for civilian research and development programs, and international science and technology policy. He is the author of *The Decision to Go to the Moon: Project Apollo and the National Interest* (1970), and articles and reports on space policy and science and technology policy. Dr. Logsdon has served as a consultant to the United Nations, the National Science Foundation, the Department of Commerce, the Office of Technology Assessment, the Environmental Protection Agency, and other public and private organizations. He is a former chairman of the Committee on Science and Public Policy of the American Association for the Advancement of Science and is currently a member of the Public Policy Committee of the American Institute for Astronautics and Aeronautics.

Pamela E. Mack is a doctoral candidate at the University of Pennsylvania, where she is completing a dissertation under Thomas P. Hughes on "The Politics of Technological Change: The History of Landsat." In 1980-1981, she held a Guggenheim Fellowship at the National Air and Space Museum. She has taught at Worcester Polytechnic Institute and Hampshire College. Her "Space Science for Applications: The History of Landsat" appeared in Paule A. Hanle and Von Del Chamberlain, eds., *Space Science Comes of Age: Perspectives in the History of the Space Sciences* (1981).

Bruce Mazlish is Professor of History at the Massachusetts Institute of Technology, where he has served as Chairman of the History Section and Head of the Humanities Department. A Fellow of the American Academy of Arts and Sciences and Associate Editor of *The Journal of Interdisciplinary History,* Professor Mazlish is a prolific author whose long list of publications begins before he took his Ph.D. from Columbia University in 1955. Perhaps best known for *The Western Intellectual Tradition* (with Jacob Bronowski) and numerous books and articles in psychohistory, Professor Mazlish is also editor of *The Railroad and the Space Program: An Exploration in Historical Analysis* (1965).

Walter A. McDougall is Associate Professor of History at the University of California, Berkeley, where he has taught since 1975. After service in Vietnam with the U.S. Army, Professor McDougall took a Ph.D. in modern European history at the University of Chicago in 1974. His revised dissertation, *France's Rhineland Diplomacy 1914-1924: The Last Bid for a Balance of Power in Europe* reinterpreted European diplomatic history between the world wars on the basis of recently opened French Foreign Ministry files. For several years Professor McDougall has been working on the international ramifications of the space program. His "Technocracy and Statecraft in the Space Age—Toward the History of a Saltation," *American Historical Review* (October, 1982), is a preview of a forthcoming, book-length study.

James A. Michener is one of America's most prolific and most read authors. Equally capable at both fiction and nonfiction, Mr. Michener is author of 30 books ranging from his dramatic and influential *Kent State: What Happened and Why* (1971) to his Pulitzer-prize-winning *Tales of the South Pacific* (1947). His long string of best-selling novels has recently been completed by *Space* (1982).

Carroll W. Pursell, Jr., is Professor of History at the University of California, Santa Barbara. Since taking his Ph.D. from Berkeley in 1962, Professor Pursell has taught at Case Institute of Technology, the University of California, Santa Barbara, and Lehigh University, where he was Andrew W. Mellon Distinguished Professor of the Humanities. Coeditor (with Melvin Kranzberg) of the standard two-volume study *Technology in Western Civilization,* Professor Pursell has written or edited numerous other works, including *Early Stationary Steam Engines in America* (1969) and most recently *Technology in America (1981).* Since 1975, he has been Secretary of the Society for the History of Technology.

John A. Simpson is Arthur H. Compton Distinguished Service Professor of Physics, University of Chicago, and Director of the Enrico Fermi Institute of Nuclear Studies. Upon taking his Ph.D. in physics from New York University in 1942, Professor Simpson joined the Manhattan Project at the University of Chicago, where he has been ever since. A member of the National Academy of Sciences, Professor Simpson established the Laboratory for Astrophysics and Space Research in 1964. He has researched and written extensively on particles and fields in space.

John Noble Wilford was educated at the University of Tennessee, Syracuse University, and Columbia University. He has been an editor or reporter since 1956, serving successively on *The Wall Street Journal* and *The New York Times.* Since 1965, he has concentrated on space exploration. His many awards include the Aviation/Space Writers Association Book Award (1970) and the Press Award of the National Space Club (1974). His books include *We Reach the Moon* (1969) and (with William Stockton) *Spaceliner* (1981).

☆U.S. GOVERNMENT PRINTING OFFICE: 455-222-1985